Building Recommender Systems Using Large Language Models

Jianqiang (Jay) Wang

Building Recommender Systems Using Large Language Models

 Springer

Jianqiang (Jay) Wang
Curify-AI
Foster City, CA, USA

ISBN 978-3-032-01151-0   ISBN 978-3-032-01152-7 (eBook)
https://doi.org/10.1007/978-3-032-01152-7

© The Editor(s) (if applicable) and The Author(s), under exclusive license to Springer Nature Switzerland AG 2025

This work is subject to copyright. All rights are solely and exclusively licensed by the Publisher, whether the whole or part of the material is concerned, specifically the rights of translation, reprinting, reuse of illustrations, recitation, broadcasting, reproduction on microfilms or in any other physical way, and transmission or information storage and retrieval, electronic adaptation, computer software, or by similar or dissimilar methodology now known or hereafter developed.
The use of general descriptive names, registered names, trademarks, service marks, etc. in this publication does not imply, even in the absence of a specific statement, that such names are exempt from the relevant protective laws and regulations and therefore free for general use.
The publisher, the authors and the editors are safe to assume that the advice and information in this book are believed to be true and accurate at the date of publication. Neither the publisher nor the authors or the editors give a warranty, expressed or implied, with respect to the material contained herein or for any errors or omissions that may have been made. The publisher remains neutral with regard to jurisdictional claims in published maps and institutional affiliations.

This Springer imprint is published by the registered company Springer Nature Switzerland AG
The registered company address is: Gewerbestrasse 11, 6330 Cham, Switzerland

If disposing of this product, please recycle the paper.

Foreword

Over the past few years, the rise of large language models (LLMs) has marked a paradigm shift not only in natural language understanding but also in how we think about personalization and recommendation.

This book, *Building Recommender Systems with Large Language Models* captures the field at an inflection point: traditional recommender models like matrix factorization and neural collaborative filtering are meeting their generative counterparts—capable of understanding and producing natural language, multimodal content, and even reasoning over structured and unstructured data. With the increasing capabilities of models like GPT series, Claude, and open-source alternatives, recommendation is no longer limited to retrieving predefined options; new systems are needed that are capable of *generation, alignment,* and *reasoning.*

Why LLMs for Recommendation Matter Today

The motivations for applying LLMs to recommendation tasks are both theoretical and practical. On one hand, LLMs offer a flexible, unified architecture that can represent user interests, item content, temporal sequences, and even conversational context without hand-crafted features or rigid schemas. On the other hand, they enable new application modes: chat-based recommendation, cold-start reasoning, dynamic personalization, and explainability.

What This Book Offers

Currently, not many resources walk readers through *how* to practically understand, build, and evaluate LLM-powered recommender systems. This book fills a crucial gap between LLM literature and recommendation system practice. It not only explains concepts with clarity and concrete data/code examples, but also uses full-length tutorials as mini-projects to run experiments. The book also discusses design trade-offs when implementing LLM-powered recommendation systems. Researchers will find a springboard for exploration. Industry practitioners will find a roadmap for deployment. And students with a background of machine learning, NLP, or data science will find an accessible and rigorous guide to this fast-moving intersection.

Why Now Is the Right Time

Now is a critical moment to engage with this space. We have seen exponential gains in generative model capabilities and also growing awareness of their limitations: latency, cost, safety, and evaluation challenges. As LLMs become more available and customizable via APIs, open-source models, and fine-tuning techniques, the field needs informed builders who can move beyond hype and toward grounded, impactful systems.

Whether you are a student, a researcher, or an engineer, this book will help you understand *why* LLMs matter for recommendation, *how* to apply them effectively and efficiently, and *what* challenges lie ahead. I'm excited for the readers of this book and for the systems they will likely build inspired by this book.

Computer Science
University of California San Diego,
La Jolla, CA, USA
                        Julian McAuley

Preface

Why This Book Was Written

Recommendation systems are at the core of modern digital experiences, from suggesting movies on streaming platforms to ranking products in e-commerce and recommending content on social media. These systems help users navigate vast amounts of information by surfacing items that match their preferences and intent.

Traditional recommendation methods emerged over time, such as collaborative filtering, matrix factorization, and deep learning models. However, they often struggle with several critical limitations: handling unstructured data, modeling complex user intent, and reasoning over sparse interactions.

Recent advances in Large Language Models (LLMs) have demonstrated powerful capabilities in language understanding, generation, reasoning, and knowledge synthesis. These strengths align closely with the growing demands of modern recommendation systems. This book explores the intersection of LLMs and recommender systems, driven by several key motivations:

- **Addressing limitations of traditional approaches**: Traditional methods lack the capacity to interpret nuanced natural language, perform complex reasoning over user intent, or effectively incorporate multi-modal and contextual information.
- **Showcasing the power of LLMs in recommendation**: LLMs offer powerful tools such as rich embeddings, few-shot prompting, reasoning over content and user profiles, and generative capabilities. These features significantly extend the capabilities of modern recommender systems.
- **Bridging research and application**: The rapid pace of development in both LLMs and recommendation systems has led to fragmented knowledge. This book aims to consolidate research insights into a structured guide, enabling practitioners to design and deploy cutting-edge systems.
- **Providing a timely, practical resource**: The intersection of LLMs and recommender systems is still emerging. This book offers a comprehensive, practice-oriented introduction to the topic, grounded in research yet oriented toward real-world applications.

Who This Book Is For

This book is intended for professionals, researchers, and students who are interested in understanding and building modern recommendation systems enhanced by Large Language Models (LLMs). Readers will benefit most if they have a foundational understanding of machine learning and natural language processing though much of the material is self-contained and accessible to those with technical curiosity.

Primary audiences include:

- Practicing data scientists, machine learning engineers, and developers working on recommendation systems or personalization.
- Graduate students and researchers in fields such as NLP, IR, AI, and data science.
- Lecturers, educators, and technical managers seeking a comprehensive resource on this rapidly evolving domain.

Recommended prerequisites:

- Basic knowledge of machine learning and NLP concepts.
- Familiarity with Python programming and frameworks like PyTorch.
- Exposure to tools such as the OpenAI API, LangChain, Hugging Face Transformers, or vector databases like Weaviate or FAISS is helpful but not mandatory.

What This Book Covers

This book is structured to provide a progressive understanding of how Large Language Models (LLMs) can be integrated into recommendation systems, from foundational concepts to advanced applications.

- **Chapter 1: Introduction to LLMs**
  Offers a foundational overview of LLMs, from tokenization and transformers to fine-tuning and inference techniques. Includes hands-on tutorials to ground the reader in core LLM concepts.
- **Chapter 2: From Traditional to LLM-Powered Recommendation Systems**
  Traces the evolution from collaborative filtering and matrix factorization to LLM-driven approaches. Introduces two paradigms of LLM-powered recommendation systems: LLM as an enhancer and LLM as recommender. Uses MovieLens data to explain the transition.
- **Chapter 3: LLM-Enhanced Recommendation Systems**
  Explores how LLMs can augment existing components such as tokenization, embedding generation, retrieval, data labeling, and evaluation. Introduces techniques like LLM-as-a-Judge and hybrid retrieval.
- **Chapter 4: LLMs as Recommender End-to-End Workflow**
  Explains how LLMs can be used directly as the recommendation engine. Covers prompting strategies, model fine-tuning, and cost-effective production deployments.

- **Chapter 5: Conversational Recommendation Systems**
  Focuses on building interactive agents that recommend through dialogue. Introduces reinforcement learning, dialogue state tracking, and clarification mechanisms, along with a hands-on product recommendation tutorial.
- **Chapter 6: Leveraging Multi-Modal Data**
  Discusses integrating multi-modal data (including text, images, audio, video) into recommendation systems. Explains the choice of multi-modal integration and multi-modal modeling, supported by a fashion recommendation case study.
- **Chapter 7: Generative Recommendation and Planning Systems**
  Explores how LLMs enable generative applications across modalities, including text, images, audio, and video. Covers techniques for generating personalized content and planning recommendation sequences. Tutorials include image-to-avatar generation and stepwise planning for goal-oriented recommendations.
- **Chapter 8: Challenges and Trends in LLMs for Recommendation Systems**
  Concludes with emerging frontiers and open questions including multi-modal integration, multi-agent systems, privacy, fairness, and verification. This chapter provides a forward-looking perspective for research and application.

Foster City, CA, USA   Jianqiang (Jay) Wang

Competing Interests The author has no competing interests to declare that are relevant to the content of this manuscript.

Contents

1	**Introduction to LLMs**		**1**
1.1	A Brief History of NLP and LLMs		1
	1.1.1	The AI Boom and Early Natural Language Processors: 1950s–1980s	2
	1.1.2	Expert Systems and Statistical Models: 1980s–1990s	2
	1.1.3	Neural Network Models, Word Embeddings, and Transformers: 2000s–2020s	3
	1.1.4	The Age of Large Language Models (LLMs): 2020–Present	3
1.2	Tokenization		4
	1.2.1	The Tokenization Workflow	4
	1.2.2	Tokenization Methods	5
		1.2.2.1 Tokenization in Language Models	6
	1.2.3	Challenges in Tokenization	6
	1.2.4	Tokenization in LLM-Powered Recommendation Systems	7
1.3	Embedding		7
	1.3.1	Types of Embeddings	8
	1.3.2	Embeddings in LLM-Powered Recommendation Systems	9
1.4	Retrieval		10
	1.4.1	The Retrieval Process	11
	1.4.2	Modern Retrieval Systems	12
		1.4.2.1 Sparse Retrieval Systems	13
		1.4.2.2 Traditional Databases with Vector Retrieval	13
		1.4.2.3 Modern Vector Databases	13
	1.4.3	Retrieval in LLM-Powered Recommendation Systems	14

1.5	Encode-Decoder and Transformer Architecture		14
	1.5.1	Encoder-Decoder Architecture	14
	1.5.2	Transformer Architecture	16
		1.5.2.1 Model Workflow	17
		1.5.2.2 Self-Attention and Q, K, V Mechanism	18
		1.5.2.3 Positional Encoding	19
		1.5.2.4 Categorization of Transformer Models	19
	1.5.3	Transformers in LLM-Powered Recommendation Systems	20
1.6	LLM Essentials	20
	1.6.1	Scale and Core Capabilities	20
	1.6.2	Emergent Abilities	21
		1.6.2.1 In-Context Learning	21
		1.6.2.2 Instruction Following	23
		1.6.2.3 Chain-of-Thought (CoT) Reasoning	23
1.7	LLM Pre-training, Post-training, and Inference	24
	1.7.1	Pre-training	24
	1.7.2	Supervised Fine-Tuning	25
		1.7.2.1 SFT Workflow	26
		1.7.2.2 Existing Frameworks for SFT	27
	1.7.3	Reinforcement Learning with Human Feedback	29
	1.7.4	LLM Inference	29
		1.7.4.1 Auto-Regressive and Speculative Decoding	30
		1.7.4.2 Architecture-Specific Inference	30
		1.7.4.3 Batching and Caching	30
		1.7.4.4 Quantization and Model Compression	30
1.8	Tutorial: Understanding Tokenization and Transformer Model	31
	1.8.1	Overview	31
	1.8.2	Experimental Design	31
	1.8.3	Results and Analysis	32
		1.8.3.1 Advanced Methods	34
	1.8.4	Conclusion	34
1.9	Second Tutorial: Understanding Content Embedding and Retrieval	35
	1.9.1	Overview	35
	1.9.2	Experimental Design	35
	1.9.3	Results and Analysis	36
		1.9.3.1 t-SNE Visualization	36
		1.9.3.2 Results Table	36
	1.9.4	Conclusion	38
References	38

2	**From Traditional to LLM-Powered Recommendation Systems**			**41**
	2.1	Recommendation System Workflow		41
		2.1.1	Content Understanding	42
			2.1.1.1 Content Understanding Tasks	42
			2.1.1.2 Content Understanding Methods	44
		2.1.2	User Modeling	45
			2.1.2.1 Collaborative Filtering (CF)	47
			2.1.2.2 Matrix Factorization (MF)	47
			2.1.2.3 Factorization Machines (FM)	48
		2.1.3	Candidate Retrieval	48
			2.1.3.1 Content-Based Retrieval	49
			2.1.3.2 Collaborative Filtering Retrieval	49
			2.1.3.3 Neural Retrieval	50
		2.1.4	Ranking	51
			2.1.4.1 Traditional Ranking Methods	52
			2.1.4.2 Traditional Reranking Methods	53
		2.1.5	Evaluation	55
			2.1.5.1 Business Metrics	55
			2.1.5.2 Model Metrics	56
			2.1.5.3 Outcome Metrics	57
	2.2	Challenges and Transition to LLM-Powered Systems		58
		2.2.1	User-Level Challenges	59
		2.2.2	Item-Level Challenges	59
		2.2.3	Model-Level Challenges	60
		2.2.4	Other Challenges and LLM Opportunities	60
	2.3	LLMs Paradigms in Recommendation Systems		61
		2.3.1	LLM-Enhanced Recommendation Systems	61
		2.3.2	LLM as Recommendation Systems	62
		2.3.3	Practical Considerations	63
	2.4	Tutorial: From Traditional to LLM-Based Recommendations Using MovieLens Dataset		64
		2.4.1	Overview	64
		2.4.2	Experimental Design	64
			2.4.2.1 Dataset and Train-Test Split	64
			2.4.2.2 Methods Compared	64
			2.4.2.3 Prompt Design	65
			2.4.2.4 Inference Model	65
			2.4.2.5 Evaluation Metrics	66
		2.4.3	Results and Analysis	66
			2.4.3.1 Results Summary	66
			2.4.3.2 Advanced Methods	67
		2.4.4	Conclusions	68
	References			68

3 LLM-Enhanced Recommendation Systems ... 71
- 3.1 Overview ... 71
- 3.2 LLM Tokenization for Recommendations ... 72
  - 3.2.1 LLM Tokenization Workflow ... 73
    - 3.2.1.1 Text Data ... 73
    - 3.2.1.2 Categorical Features ... 74
    - 3.2.1.3 Numerical Features ... 74
    - 3.2.1.4 Multi-modal Data ... 74
  - 3.2.2 Integrating LLM Tokenization to Recommendation Systems ... 75
    - 3.2.2.1 Semantic Tokenization and Concept Extraction ... 75
    - 3.2.2.2 Hybrid Modeling with Semantic Tokens ... 75
    - 3.2.2.3 Dynamic Trend Adaptation ... 76
- 3.3 Embeddings from Unstructured Data ... 76
  - 3.3.1 Obtaining LLM Embeddings ... 76
  - 3.3.2 Storing Embeddings ... 77
  - 3.3.3 Evaluating Embeddings ... 78
    - 3.3.3.1 Retrieval Quality ... 78
    - 3.3.3.2 Labeled Similarity Data ... 78
    - 3.3.3.3 Downstream Task Performance ... 79
- 3.4 LLM-Augmented Retrieval ... 79
  - 3.4.1 Dense Retrieval ... 80
    - 3.4.1.1 Locality-Sensitive Hashing ... 80
    - 3.4.1.2 Space-Partitioning Algorithms (e.g., KD-Trees, Annoy) ... 81
    - 3.4.1.3 Graph-Based Traversal Algorithms (e.g., NSW, HNSW) ... 82
  - 3.4.2 Industrial Tools for Dense Retrieval ... 83
    - 3.4.2.1 FAISS (Facebook AI Similarity Search) ... 83
    - 3.4.2.2 ScaNN (Scalable Nearest Neighbors by Google) ... 83
  - 3.4.3 LLM-Enhanced Retrieval ... 84
    - 3.4.3.1 Query Rewriting ... 84
    - 3.4.3.2 Contextual Augmentation ... 84
- 3.5 LLM-Based Data Labeling and Evaluation ... 85
  - 3.5.1 LLM-as-a-Judge for Recommendation Evaluation ... 85
    - 3.5.1.1 Key Frameworks ... 85
    - 3.5.1.2 General Workflow ... 85
    - 3.5.1.3 Two Approaches ... 86
  - 3.5.2 Human-Assisted LLM Labeling ... 87
- 3.6 Tutorial: Topic Classification and Item Similarity Labeling Using LLMs ... 88
  - 3.6.1 Overview ... 88
  - 3.6.2 Experimental Design ... 88
    - 3.6.2.1 Dataset Setup ... 89

		3.6.2.2	LLM Choices	89
		3.6.2.3	Labeling Methods	89
		3.6.2.4	Evaluation Metrics	90
	3.6.3	Results and Analysis		90
		3.6.3.1	Topic Classification Labeling	90
		3.6.3.2	Item Similarity Labeling	92
	3.6.4	Conclusions		95
3.7	Tutorial: News Recommendation by Combining Embedding with Learning-to-Rank Models			95
	3.7.1	Overview		95
		3.7.1.1	Goal of the Tutorial	95
	3.7.2	Experimental Design		95
		3.7.2.1	Data	95
		3.7.2.2	Retrieval Set Generation	96
		3.7.2.3	Ground Truth Labeling	96
		3.7.2.4	Recommendation Approaches	96
		3.7.2.5	Evaluation	96
	3.7.3	Results and Analysis		97
	3.7.4	Conclusion		98
References				98

4 LLM as Recommender 99
4.1 LLMs as Recommender End-to-End Workflow 99
4.1.1 Step 1: Input Data Preparation 100
4.1.2 Step 2: Prompt Engine 102
4.1.3 Step 3: LLM Inference 102
4.1.4 Step 4: Post-Processing (Optional) 103
4.1.5 Step 5: Evaluation 103
4.2 Prompting for Recommendation 103
4.2.1 Prompting Techniques 104
4.2.1.1 Zero-Shot Prompting 104
4.2.1.2 Few-Shot Prompting 104
4.2.1.3 Instruction-Based Prompting 105
4.2.1.4 Chain-of-Thought (CoT) Prompting 105
4.2.2 Prompting for Various Recommendation Tasks ... 106
4.2.3 Prompt Design Practical Tips 107
4.3 Fine-Tuning LLMs for Recommendation 107
4.3.1 Instruction Fine-Tuning for Recommendation 108
4.3.1.1 Task Types 109
4.3.1.2 Benefits 110
4.3.2 Domain Knowledge Fine-Tuning 110
4.3.2.1 Implementation and Examples 110
4.3.2.2 Benefits 111
4.3.3 Personalized LLM Fine-Tuning 111
4.3.3.1 Implementation 111

		4.3.3.2 Choice Between Personalized User Embedding and Personalized LLM Fine-Tuning	112
	4.3.4	Summary and Discussion............................	113
4.4	Production-Ready Optimization for LLM as Recommender......		115
	4.4.1	Knowledge Distillation	115
		4.4.1.1 Knowledge Distillation Implementation..........	116
		4.4.1.2 Benefits	116
		4.4.1.3 Challenges.................................	116
		4.4.1.4 Best Practices and Considerations	117
	4.4.2	Quantization and Model Compression..................	117
		4.4.2.1 Techniques and Software Packages..............	118
	4.4.3	Caching and Response Reuse	118
		4.4.3.1 Benefits	119
		4.4.3.2 Limitations	119
		4.4.3.3 Caching Strategy and Best Practice	119
	4.4.4	Design Trade-Offs and Practical Considerations.........	120
		4.4.4.1 Cost Vs. Quality Vs. Latency	121
		4.4.4.2 Fine-Tuning Vs. Retrieval-Augmented Generation (RAG)...........................	121
		4.4.4.3 Self-Built Models Vs. APIs....................	121
		4.4.4.4 Prompt-Driven Versatility and Minimizing Dependencies	122
		4.4.4.5 Scalability and Cost-Efficiency.................	122
4.5	Tutorial: Fine-Tuning LLMs for Personalized Movie Recommendations		122
	4.5.1	Overview	122
	4.5.2	Experimental Design	123
		4.5.2.1 Dataset Preparation..........................	123
		4.5.2.2 Prompt Construction.........................	123
		4.5.2.3 Fine-Tuning Approach	124
		4.5.2.4 Evaluation	124
	4.5.3	Results and Analysis	124
	4.5.4	Conclusion......................................	125
		4.5.4.1 Recommendations...........................	125
		4.5.4.2 Key Takeaways	125
4.6	Second Tutorial: Knowledge Distillation Using MovieLens Dataset		125
	4.6.1	Overview	125
	4.6.2	Experimental Design	126
		4.6.2.1 Dataset Preparation..........................	126
		4.6.2.2 Teacher and Student Models...................	127
		4.6.2.3 Distillation Process..........................	127
		4.6.2.4 Evaluation Metrics	127
	4.6.3	Results and Analysis	127
4.7	Conclusions		128
References...			129

5	**Conversational Recommendation Systems**		131
	5.1	Reinforcement Learning Foundations for Conversational Recommendation	131
		5.1.1 Introduction	131
		5.1.2 Types of RL Algorithms in Recommendation	132
		5.1.2.1 Multi-Armed Bandit (MAB)	132
		5.1.2.2 Deep Q-Networks (DQN)	133
		5.1.2.3 Policy Gradient Methods	133
		5.1.2.4 Monte Carlo Tree Search (MCTS)	134
		5.1.3 Integrating RL with LLMs in Conversational Recommendation	135
		5.1.3.1 Roles of LLM and RL	135
		5.1.3.2 Dialogue-Level Reward Design	135
		5.1.3.3 Pipeline Integration Strategy	136
		5.1.3.4 Benefits of LLM-RL Integration	136
	5.2	Key Modules in CRS	136
		5.2.1 Dialogue and Intent Management	137
		5.2.1.1 Intent Detection	138
		5.2.1.2 Slot Filling	139
		5.2.1.3 Dialogue State Tracking (DST)	140
		5.2.2 Clarification and Feedback Mechanisms	141
		5.2.3 Personalization and Context Handling	142
		5.2.3.1 Context Handling	142
		5.2.3.2 Personalization	142
		5.2.4 Continuous Evaluation	143
		5.2.4.1 Evaluation Data Sources	143
		5.2.4.2 Evaluation Metrics	144
		5.2.5 Reward Design in CRS	144
		5.2.5.1 Example: Reward Computation for a Dialogue Turn	144
	5.3	Designing Conversational Recommender Systems	145
		5.3.1 System Architecture and Workflow Integration	145
		5.3.2 Data and Infrastructure Requirements	145
		5.3.3 Performance Optimization and Iterative Improvement	146
		5.3.3.1 System Optimization Techniques	147
		5.3.3.2 Performance Tracking and Iterative Refinement	147
	5.4	Tutorial: Conversational Recommendation System with RL and LLMs	148
		5.4.1 Overview	148
		5.4.2 Experimental Design	148
		5.4.2.1 Dataset Design	148
		5.4.2.2 Methodology	149

		5.4.3	Results and Analysis	151
			5.4.3.1 User Preference Extraction	151
			5.4.3.2 Reward Dynamics	151
			5.4.3.3 Discussions	152
	5.5	Conclusion		153
	References			153
6	**Leveraging Multi-modal Data**			**155**
	6.1	Introduction		155
		6.1.1	Core Modalities and Their Roles	155
		6.1.2	The Multi-modal Advantage	156
		6.1.3	Challenges in Multi-modal Integration	157
		6.1.4	Modeling Strategies	157
	6.2	Multi-modal Integration Techniques		157
		6.2.1	Early Fusion	159
		6.2.2	Late Fusion	159
		6.2.3	Hybrid Fusion	160
	6.3	Multi-modal LLMs		160
		6.3.1	Modeling Principles of Multi-modal LLMs	161
			6.3.1.1 Tokenization and Modality Encoding	161
			6.3.1.2 Fusion Strategies	161
			6.3.1.3 Cross-Modal Alignment Objectives	161
			6.3.1.4 From Principles to Model Designs	162
		6.3.2	Advantages and Limitations	162
		6.3.3	Choice Between Multi-modal Integrations and Multi-modal LLMs	163
	6.4	Tutorial: Multi-modal Fashion Recommendation with Pairwise Ranking		164
		6.4.1	Overview	164
		6.4.2	Experimental Design	165
		6.4.3	Results and Analysis	166
	6.5	Conclusion		166
	References			167
7	**Generative Recommendation and Planning Systems**			**169**
	7.1	Introduction		169
		7.1.1	Motivations	170
		7.1.2	Content Generation Summary	170
		7.1.3	Text Generation	171
			7.1.3.1 Categorization by Input Types	171
			7.1.3.2 Modeling Architectures	171
			7.1.3.3 Controlled Generation Techniques	172
		7.1.4	Image Generation	172
			7.1.4.1 Categorization by Input Types	172
			7.1.4.2 Modeling Architectures	173
			7.1.4.3 Controlled Generation	173

		7.1.5	Audio Generation	174

- 7.1.5 Audio Generation ... 174
  - 7.1.5.1 Categorization by Input Types ... 174
  - 7.1.5.2 Modeling Architectures ... 175
  - 7.1.5.3 Controlled Generation ... 175
  - 7.1.5.4 Example: TTS with Bark (Suno AI) ... 176
- 7.1.6 Video Generation ... 177
  - 7.1.6.1 Categorization of Input Types ... 177
  - 7.1.6.2 Modeling Architectures ... 177
  - 7.1.6.3 Controlled Generation Techniques ... 178
  - 7.1.6.4 Script-to-Video Example: Intelligent Museum Narrative ... 178

7.2 Evaluation ... 181
- 7.2.1 Constructing Benchmark Data ... 182
- 7.2.2 Dimensions and Metrics ... 182
  - 7.2.2.1 Fidelity and Quality ... 183
  - 7.2.2.2 Relevance and Alignment ... 184
  - 7.2.2.3 Diversity and Expressiveness ... 184
  - 7.2.2.4 Safety and Toxicity ... 184
- 7.2.3 Evaluation Method ... 185

7.3 Sequential Planning with LLMs ... 185
- 7.3.1 Key Components ... 186
  - 7.3.1.1 Sequential Decision-Making ... 186
  - 7.3.1.2 Planning as Constrained Generation ... 186
  - 7.3.1.3 Dynamic Adaptation ... 186
  - 7.3.1.4 Verifiable Outcomes ... 187
- 7.3.2 Application Scenarios ... 187
  - 7.3.2.1 Example of External Verification ... 188

7.4 Tutorial: Image-to-Avatar Generation ... 188
- 7.4.1 Overview ... 188
- 7.4.2 Experimental Design ... 189
  - 7.4.2.1 Data Source ... 189
  - 7.4.2.2 Methods ... 189
  - 7.4.2.3 Evaluation Metrics ... 190
- 7.4.3 Results and Analysis ... 190
- 7.4.4 Discussion ... 191

7.5 Second Tutorial: Goal-Driven Planning with LLMs ... 191
- 7.5.1 Overview ... 191
- 7.5.2 Experimental Design ... 192
  - 7.5.2.1 User Constraint Specification ... 192
- 7.5.3 Results and Analysis ... 194
  - 7.5.3.1 Generated Plan (Excerpt) ... 194
  - 7.5.3.2 Tool-Based Verification ... 194
  - 7.5.3.3 LLM-as-a-Judge Evaluation ... 194
- 7.5.4 Discussion ... 195

References ... 195

8 Challenges and Trends in LLMs for Recommendation Systems 197
8.1 Introduction 197
8.2 Multi-modal Integration 198
8.2.1 Challenges 198
8.2.1.1 Data Alignment 198
8.2.1.2 Consistency Across Modalities 198
8.2.1.3 Computational Complexity 199
8.2.2 Promising Directions 199
8.2.2.1 Cross-Modal Pre-training 199
8.2.2.2 Efficient Fusion Techniques 199
8.2.2.3 User-Centric Multi-modal Interfaces 200
8.3 Verifiable Outcomes 200
8.3.1 Challenges 201
8.3.1.1 Delayed or Ambiguous Feedback 201
8.3.1.2 Interpretability and Justification 201
8.3.1.3 Data Provenance and Credibility 201
8.3.2 Promising Directions 201
8.3.2.1 Reasoning LLMs 201
8.3.2.2 Interactive Explanations 202
8.4 Multi-agent Systems 202
8.4.1 Challenges 202
8.4.1.1 Agent Collaboration and Error Propagation 202
8.4.1.2 Human-in-the-Loop Complexity 203
8.4.1.3 Scalability and Maintenance 203
8.4.2 Promising Directions 203
8.4.2.1 Agent Framework Innovations 203
8.4.2.2 Simulated Environments 203
8.4.2.3 Hybrid Human-Agent Systems 204
8.5 Generative Copyright and Privacy 204
8.5.1 Challenges 205
8.5.1.1 Copyright Infringement 205
8.5.1.2 Data Privacy 205
8.5.1.3 Legal Ambiguity 205
8.5.2 Promising Directions 205
8.5.2.1 Synthetic Data Generation 205
8.5.2.2 Differential Privacy and On-Device Personalization 205
8.6 Ethical AI and Fairness 206
8.6.1 Challenges 206
8.6.1.1 Bias in Training Data 206
8.6.1.2 Defining Fairness 206
8.6.1.3 Transparency 206

		8.6.2	Promising Directions.............................	207
			8.6.2.1 Bias Detection and Mitigation	207
			8.6.2.2 Explainable AI (XAI)	207
8.7	Conclusions			207
References..				207

Index.. 209

Chapter 1
Introduction to LLMs

This chapter introduces the key ideas behind Large Language Models (LLMs) and their growing role in AI-powered recommendation systems. It reviews the development of natural language processing (NLP), from early rule-based methods to modern transformer architectures like BERT and GPT, providing the foundation for understanding how LLMs process unstructured text. Core concepts are covered, such as tokenization, embeddings, attention mechanisms, and retrieval techniques. The chapter also explains how LLMs are built and trained, including pre-training, fine-tuning, reinforcement learning with human feedback (RLHF). To connect theory with practice, the chapter provides hands-on tutorials that guide readers through tasks like tokenization, attention visualization, and retrieval pipelines—empowering both new learners and experienced users to apply LLMs to real-world challenges.

1.1 A Brief History of NLP and LLMs

To explore the role of Large Language Models (LLMs) in recommendation systems, it is essential to recognize that these models are the culmination of decades of interdisciplinary progress. Advances in linguistics, computer science, mathematics, statistics, and hardware technology have collectively shaped the development of LLMs. Understanding their impact requires a historical perspective, tracing the evolution of Natural Language Processing (NLP) from early rule-based systems to the sophisticated models of today. This journey through NLP's milestones highlights how each breakthrough has paved the way for modern recommendation systems. Figure 1.1 illustrates this evolutionary trajectory, spanning from the 1950s to the present.

© The Author(s), under exclusive license to Springer Nature
Switzerland AG 2025
J. (J.) Wang, *Building Recommender Systems Using Large Language Models*,
https://doi.org/10.1007/978-3-032-01152-7_1

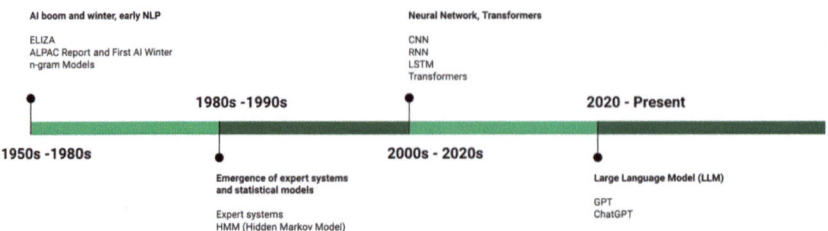

Fig. 1.1 A Brief timeline of the evolution of NLP from the 1950s to the present state

1.1.1 The AI Boom and Early Natural Language Processors: 1950s–1980s

The first phase began in the 1950s with the enthusiasm for artificial intelligence (AI). The early history of AI experienced several peaks, followed by periods of disillusionment due to unmet high expectations, leading to the "AI winter." Notable achievements during this phase included advancements in *linguistic theory, early prototypes of dialogue systems,* and the application of *n-gram models*.

In linguistic theory, a key milestone was Noam Chomsky's *Syntactic Structures* (Chomsky, 1957), which introduced generative grammar and provided a theoretical foundation for parsing and machine translation. The 1970s saw further innovations in computational semantics, such as case grammar, semantic networks, and conceptual dependency theory.

In parallel, rule-based dialogue systems began to emerge. A seminal example was ELIZA (Weizenbaum, 1966), which simulated a Rogerian psychotherapist using pattern-matching rules. Despite its simplicity, many users believed it understood them—marking a foundational step in human-computer interaction and conversational agents.

Another major contribution was the *n-gram language model*, rooted in Markov chains (Markov, 1913) and later popularized by Shannon's experiments in the 1940s. n-Grams became practical in the 1970s and 1980s, thanks to work at IBM (Jelinek, 1976) and CMU (Baker, 1975), powering early speech recognition systems. Despite their limitations, n-grams laid the groundwork for probabilistic language modeling.

1.1.2 Expert Systems and Statistical Models: 1980s–1990s

The second phase of NLP development saw a shift toward *expert systems* and *statistical models*. Rule-based expert systems like MYCIN used hard-coded inference rules and ontologies to process natural language. However, by the late 1980s, data-driven approaches began to dominate, thanks to increasing computational power and large corpora.

Hidden Markov Models (HMMs) became the workhorse for sequential tasks like *part-of-speech tagging, named entity recognition*, and *speech recognition*. HMMs model observable word sequences and hidden state transitions efficiently, often using the *Viterbi algorithm* for decoding.

During this time, the seeds of neural NLP were planted. Recurrent Neural Networks (RNNs) were introduced by Elman (1990), and later enhanced with Long Short-Term Memory (LSTM) networks by Hochreiter and Schmidhuber (1997) to address long-range dependencies.

1.1.3 Neural Network Models, Word Embeddings, and Transformers: 2000s–2020s

The third phase brought *neural networks* to the forefront. A breakthrough came with *Word2Vec* (Mikolov et al., 2013), which enabled unsupervised learning of high-quality *word embeddings*. These continuous representations outperformed sparse models in many tasks and formed the foundation of modern NLP.

This era also saw creative use of Convolutional Neural Networks (CNNs) for tasks like text classification (Kim, 2014), and RNNs for sequence modeling, such as machine translation. However, both had limitations in modeling long dependencies.

To address this, the *encoder-decoder architecture* (Sutskever et al., 2014) was introduced, enabling end-to-end sequence transduction. The addition of *attention mechanisms* (Bahdanau et al., 2014) improved alignment in translation tasks.

A pivotal moment arrived with the *Transformer* architecture (Vaswani et al., 2017), which replaced recurrence entirely with self-attention mechanisms. Transformers allowed efficient parallel training and better global context modeling, becoming the backbone of nearly all state-of-the-art models today.

1.1.4 The Age of Large Language Models (LLMs): 2020–Present

The advent of Transformers ushered in a transformative era for NLP. The 2020s witnessed the emergence of Large Language Models (LLMs), trained on vast datasets and capable of tasks such as text generation, language translation, question answering, and more. Models like GPT-3 (Brown et al., 2020), BERT (Devlin et al., 2018), and T5 (Raffel et al., 2020) excel at producing human-like text and capturing nuanced semantics, making them ideal for applications such as recommendation systems.

LLMs represent the pinnacle of decades of NLP advancements. Unlike earlier models constrained by rule-based or shallow statistical methods, LLMs generate dynamic, context-aware content in real time. This adaptability is particularly critical in recommendation systems, where understanding user context and delivering personalized suggestions are paramount. Additionally, the development of general-purpose embedding techniques and sequence learning architectures laid the groundwork for today's LLM-based recommenders.

1.2 Tokenization

Tokenization is a fundamental step in Natural Language Processing (NLP), involving the breakdown of text into smaller units called tokens, which can be words, subwords, or characters. These tokens are then converted into numerical representations for model processing.

This section covers the tokenization workflow, the various types of tokenization, and the challenges involved.

1.2.1 The Tokenization Workflow

Tokenization involves several steps, each of which ensures that the input text is transformed into a suitable form for LLMs to process effectively. Figure 1.2 illustrates a basic tokenization workflow.

1. **Splitting the Input into Tokens**: The first step is splitting a sentence into its smaller units, or tokens. For example:
   - Input: *"This course is amazing!"*
   - Tokens: *["This", "course", "is", "amazing", "!"]*

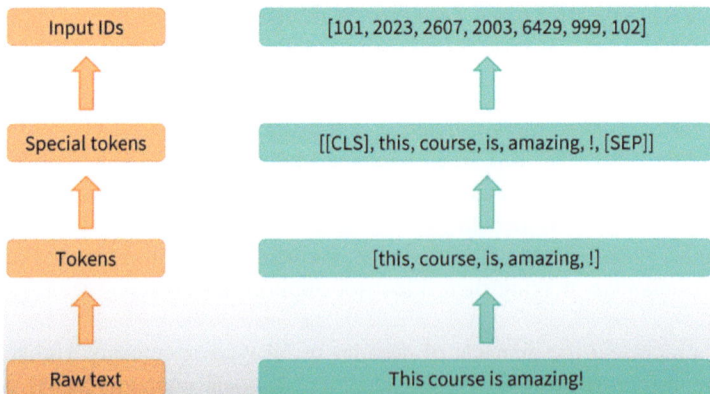

Fig. 1.2 Tokenization workflow illustrated by the sentence "This course is amazing!"

2. **Adding Special Tokens**: Special tokens are used to serve specific purposes within the model (Devlin et al., 2019). These tokens can help indicate things like sentence boundaries, masking for prediction tasks, or classification markers. Examples include:
   - **[SEP]**: A token used to separate two sentences (in tasks like *sentence-pair classification*).
   - **[MASK]**: Used in tasks where the model is asked to predict a missing word (e.g., BERT's masked language modeling).
   - **[CLS]**: A token that represents the "classification" of an entire sequence in models like BERT.
3. **Mapping Tokens to Integers**: Once the tokens are split and special tokens are added, each token is mapped to an integer ID based on a *predefined vocabulary*. This step is essential because LLMs don't understand text in its raw form—they work with *numerical representations*. For example, the token "amazing" could be mapped to the integer ID 4321, which corresponds to that word in the model's vocabulary.

1.2.2 Tokenization Methods

Tokenization is the process of breaking down text into smaller units called *tokens*. These tokens are the building blocks of language models, and the method used for tokenization can significantly influence model performance.

There are three common tokenization strategies: *word-level*, *character-level*, and *subword-level*.

- **Word-Level Tokenization** splits text into words based on spaces or punctuation.
  Example: The sentence "This course is amazing" becomes ["This", "course", "is", "amazing"].
  This approach is simple and was used in early models like *Word2Vec* (Mikolov et al., 2013).
- **Character-Level Tokenization** breaks text into individual characters.
  Example: The word "amazing" becomes ["a", "m", "a", "z", "i", "n", "g"].
  This method is used in models such as CharCNN *(*Kim et al., 2016) and is helpful for handling misspellings and unseen words.
- **Subword-Level Tokenization** divides words into smaller, meaningful units, often capturing roots and affixes.
  Example: "tokenization" might be split into "token" and "##ization".
  Popular subword algorithms include Byte Pair Encoding (BPE) (Sennrich et al., 2016), WordPiece (Schuster & Nakajima, 2012), and SentencePiece (Kudo & Richardson, 2018).

Table 1.1 summarizes the advantages and limitations of each approach:

Table 1.1 Comparison of tokenization methods

Method	Pros	Limitations
Word	– Simple and interpretable – Works well for languages with clear word boundaries	– Large vocabulary size – Struggles with out-of-vocabulary (OOV) words – Weak at capturing morphological variation
Character	– Very small vocabulary – Handles OOV and misspelled words well	– Lacks semantic meaning – Longer sequences increase complexity – Poor at modeling long-range dependencies
Subword	– Balances vocabulary size and OOV handling – Captures word similarity and morphology	– Requires careful design – May over-segment frequent words – Rare words can still pose issues

Table 1.2 Tokenization techniques used in popular LLMs

Tokenization method	Description	Example models
Character-level BPE	Merges frequently occurring character pairs into subwords	GPT
Byte-level BPE	Tokenizes raw bytes, allowing support for any character	GPT-2, RoBERTa
WordPiece	Uses statistical likelihood to build subword units	BERT, DistilBERT
SentencePiece	Learns subword units from raw text using BPE or unigram	XLNet, ALBERT

1.2.2.1 Tokenization in Language Models

Different large language models (LLMs) use different tokenization methods, often tailored to their training data and architecture. Each tokenizer is trained on a large corpus and used to preprocess input before feeding it into the model. Table 1.2 outlines tokenization techniques and their corresponding models.

1.2.3 Challenges in Tokenization

Despite its fundamental importance, tokenization presents several challenges that can impact model performance. Below are some of the most prevalent issues and their potential solutions:

1. **Out-of-Vocabulary (OOV) Tokens**
   Problem: Tokens not present in the model's vocabulary are often replaced with a special unknown token (e.g., <UNK>), leading to potential information loss.
   Solution: Subword tokenization methods, such as Byte Pair Encoding (BPE), mitigate OOV issues by decomposing words into smaller, more frequent subword units.

2. **Handling Multilingual Data**
   Problem: Global applications often require tokenizing text across multiple languages. Many tokenization schemes are language-specific, complicating multilingual processing.
   Solution: Language-agnostic tokenization techniques, such as SentencePiece, enable models to process multilingual data without requiring separate tokenizers for each language.
3. **Token Length Limitations**
   Problem: LLMs like GPT-3 and BERT impose strict token limits (e.g., 2048 tokens for GPT-3). Exceeding these limits necessitates truncation or splitting of text, which can result in context loss.
   Solution: Strategies such as sliding windows or chunking can address length constraints, though they may still compromise contextual coherence.

1.2.4 Tokenization in LLM-Powered Recommendation Systems

Tokenization serves as a critical first step in LLM-powered recommendation systems, converting raw inputs—such as content descriptions, user profile information, and user or item IDs—into structured sequences that language models can process.

- In *embedding-based retrieval*, tokenization allows textual data to be transformed into dense vector representations, making it easier to match users with relevant items.
- In *prompt-based approaches*, it ensures that complex queries and context-rich inputs are accurately interpreted, enabling the LLM to generate personalized and coherent recommendations.

By standardizing diverse input types into a unified format, tokenization supports both retrieval and generation tasks, and plays a key role in ensuring consistent and effective recommendation performance.

1.3 Embedding

Embeddings are dense, high-dimensional vectors that represent data in a way that captures meaning and relationships. Whether you're dealing with words, sentences, or entire documents, embeddings translate raw input into a form that a model can process and understand. Imagine two words, like "king" and "queen." These words will have similar embeddings, but their relationship—such as gender—is reflected in their proximity within the embedding space. By mapping complex data into these vectors, embeddings allow models to recognize similarities, differences, and contextual meanings.

Embeddings exist in a high-dimensional space. While we can't directly visualize this space due to its complexity, we can project it into two or three dimensions for easier interpretation. In such a visualization, items that are semantically similar will appear close together, while dissimilar ones will be farther apart.

1.3.1 Types of Embeddings

Embeddings play a pivotal role in LLM-based recommendation systems, with different types of embeddings offering unique capabilities tailored to specific tasks. These embeddings vary in their ability to capture syntactic, semantic, and contextual relationships, making them suitable for diverse applications.

Below, we explore the most common types of embeddings, their underlying methodologies, and their relevance to recommendation systems.

- **Word Embeddings** are vector representations of words, designed to capture both syntactic (structural) and semantic (meaning-based) relationships. These embeddings map words into a continuous vector space, where words with similar meanings or usage patterns are positioned closer together. For example, the words "king" and "queen" would have similar embeddings but differ slightly to reflect their gender-based relationship.
    - **Popular Methods**:
        Word2Vec (Mikolov et al., 2013) employs either the Continuous Bag-of-Words (CBOW) or Skip-Gram methods to predict words based on their context, focusing on local word relationships.
        GloVe (Pennington et al., 2014) leverages global word co-occurrence statistics to create word vectors, capturing broader patterns in the data.
    - **Application**: Enable fine-grained content analysis through distributional semantics, supporting:
        Cold-start item categorization by clustering product descriptions into embedding-derived taxonomies, enabling recommendations for new items without historical interactions.
        Cross-lingual recommendation by mapping user queries and item metadata into aligned multilingual embedding spaces, allowing retrieval across languages (e.g., "libro" → "book").
- **Sentence Embeddings** extend the concept of word embeddings to represent entire sentences or longer text segments. These embeddings aim to capture the overall meaning of a sentence rather than just the individual words within it. For instance, the sentences "I saw a man with a telescope" and "With a telescope, I saw a man" contain the same words but convey different meanings.
    - **Popular Methods**:
        Simple methods, such as mean pooling of the word embeddings of a sentence, can provide a basic representation but often fail to preserve nuanced information.

1.3 Embedding

InferSent (Conneau et al., 2017) uses a BiLSTM with max-pooling trained on natural language inference data, while the Universal Sentence Encoder (Cer et al., 2018) leverages either a deep averaging network or a Transformer-based model to capture sentence meaning.

Sentence-BERT (SBERT), a modification of BERT that uses siamese/triplet network structures to generate semantically meaningful sentence embeddings (Reimers & Gurevych, 2019).

- **Application**: Facilitate document-level recommendation by:

    Computing semantic similarity between user queries/reviews and item descriptions, enabling more accurate retrieval and ranking.

    Zero-shot recommendation by leveraging semantic embedding spaces to recommend relevant items in domains or languages not seen during training.

    Cross-modal alignment, such as linking text reviews to visual product embeddings, to support multimodal recommendation (e.g., recommending fashion items based on textual reviews).

- **Contextual Embeddings** represent a significant advancement in embedding technology, as they dynamically adapt to the context in which a word appears. Unlike static word embeddings, contextual embeddings generate different vector representations for the same word depending on its usage within a sentence. This capability is particularly valuable for resolving polysemy, where a single word can have multiple meanings. For example, the word "bank" would have distinct embeddings when referring to a financial institution versus the side of a river.

    - **Popular Methods**:

        ELMo (Peters et al., 2018) uses pre-trained LSTM layers to produce deep contextualized word embeddings based on entire sentence context.

        BERT (Devlin et al., 2019) adopts a bidirectional transformer architecture to learn context-aware embeddings by looking both left and right of a word.

    - **Application**: Power dynamic recommendation through:

        Session-aware sequential modeling, where contextual embeddings track evolving user interaction histories to predict next-item preferences (e.g., Transformer-based models for sequential recommendation).

        Personalized query understanding, where contextualized token-level embeddings capture user intent in natural language queries, improving retrieval and ranking accuracy.

1.3.2 Embeddings in LLM-Powered Recommendation Systems

Embeddings are central to LLM-powered recommendation systems, enabling the representation of user preferences and item attributes to uncover complex relationships and deliver personalized recommendations.

- **User Preferences**: Embeddings capture user behavior and preferences from interaction history. For example, a user interested in "budget-friendly laptops with long battery life" will generate embeddings reflecting these preferences, allowing the system to recommend similar items even without exact keyword matches.
- **Item Attributes**: Embeddings also represent item characteristics, derived from descriptions, reviews, and metadata. For instance, a product description like "lightweight laptop with long battery life" creates an embedding that highlights key attributes, helping the system match it to relevant user preferences.
- **Matching and Ranking**: Embeddings enable similarity computations between user preferences and item attributes. By comparing user and item embeddings, the system ranks items based on proximity in the embedding space, ensuring the most relevant recommendations.
- **Cold-Start Problem**: Embeddings help address the cold-start issue by using metadata (e.g., categories or tags) to generate recommendations for new users or items, even in the absence of sufficient historical data. This allows for meaningful suggestions until more personalized data becomes available.

1.4 Retrieval

Retrieval is a critical step in recommendation systems, where the goal is to identify and surface the most relevant items for a user based on their preferences or interaction history. This process typically involves comparing representations of users and items in a high-dimensional space to determine their similarity.

Retrieval methods can be broadly categorized into *dense retrieval* and *sparse retrieval*, each with its own strengths and applications. Additionally, *hybrid approaches* that combine both methods are increasingly being adopted to leverage the advantages of both paradigms.

- **Dense Retrieval**: In dense retrieval, user preferences and item attributes are represented as dense, continuous vectors (embeddings) in a high-dimensional space. These embeddings are typically generated using deep learning models, such as transformers, which capture semantic relationships and nuanced patterns. Dense retrieval excels at understanding contextual and semantic similarities, making it particularly effective for tasks where user preferences or item descriptions are complex or implicit.
- **Sparse Retrieval**: Sparse retrieval, on the other hand, relies on sparse vector representations, often derived from traditional methods like TF-IDF or BM25. These representations focus on explicit keyword matches or term frequencies, making them more interpretable and computationally efficient. Sparse retrieval is particularly useful when exact keyword matching or term-based relevance is critical, such as in scenarios where users have specific, well-defined preferences.
- **Hybrid Retrieval**: To harness the strengths of both dense and sparse retrieval, hybrid approaches are often employed. These methods combine dense and sparse representations, either by merging their similarity scores or by using dense

embeddings to refine sparse retrieval results (or vice versa). Hybrid retrieval can improve retrieval quality by balancing semantic understanding with precise keyword matching, making it particularly effective in scenarios where both implicit and explicit user preferences are important.

Example (FAISS + BM25 Hybrid)
- Use BM25 to retrieve textually relevant items.
- Use LLM embeddings to compute semantic similarity.
- Fuse the rankings (e.g., weighted average or rank aggregation).

```
# Retrieve top-k from BM25 and FAISS separately
bm25_results = retrieve_bm25(query_text)
dense_results = retrieve_faiss(llm_embedding)
# Combine using weighted score
final_ranking = weighted_fusion(bm25_results, dense_results,
alpha=0.6)
```

1.4.1 The Retrieval Process

The retrieval process in recommendation systems encompasses both dense and sparse retrieval methods and can be outlined as follows (Fig. 1.3):

1. **Generate Embedding/Representations**:
   - For dense retrieval, generate dense embeddings for user preferences (e.g., interaction history) and item attributes using deep learning models.
   - For sparse retrieval, generate sparse representations based on term frequencies or keyword matches.

2. **Store Embedding/Representations**:
   - Store item embeddings (dense or sparse) in a database or search index optimized for efficient retrieval.

3. **Compute Similarity**:
   - For dense retrieval, compute the similarity between the user's dense embedding and item embeddings using metrics like cosine similarity or dot product.
   - For sparse retrieval, compute similarity using methods like BM25 or TF-IDF scoring.

4. **Retrieve Top-k Items**:
   - Retrieve the top-k items based on the computed similarity scores, ensuring that the most relevant items are surfaced to the user.

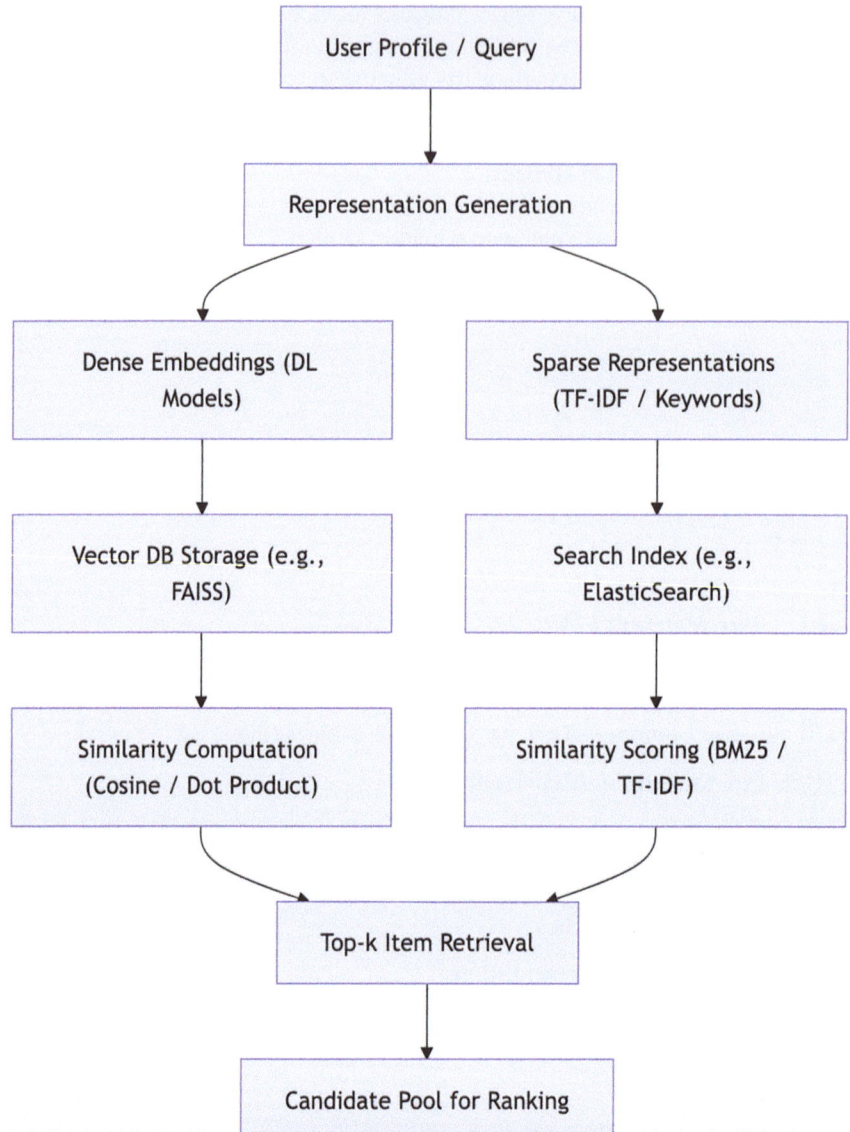

Fig. 1.3 Hybrid retrieval workflow diagram

1.4.2 Modern Retrieval Systems

Retrieval systems are essential for identifying relevant items based on user preferences. They can be categorized into *sparse retrieval* (keyword-based) and *dense retrieval* (embedding-based), with hybrid approaches combining both for enhanced

1.4 Retrieval

performance. Hybrid retrieval combines dense and sparse methods to leverage both semantic understanding and keyword matching. Tools like Elasticsearch and Weaviate natively support hybrid retrieval, enabling more comprehensive and accurate recommendations.

1.4.2.1 Sparse Retrieval Systems

Sparse retrieval focuses on explicit keyword matching, making it ideal for scenarios requiring interpretability and precision. Key systems include:

1. **Elasticsearch**: A distributed search engine optimized for sparse retrieval using TF-IDF or BM25 scoring. It supports advanced features like filtering and faceting.
2. **Apache Solr**: Similar to Elasticsearch, Solr offers flexible and extensible sparse retrieval capabilities.
3. **OpenSearch**: A community-driven alternative to Elasticsearch, providing robust sparse retrieval functionality.

These systems are widely used in e-commerce and content recommendation, where keyword-based matching is critical.

1.4.2.2 Traditional Databases with Vector Retrieval

Traditional relational databases have adapted to support vector retrieval, enabling hybrid recommendation systems:

1. **PostgreSQL with** pgvector: The pgvector extension allows PostgreSQL to store and query high-dimensional vectors, supporting similarity search metrics like cosine similarity and L2 distance.
2. **MySQL**: While still evolving, MySQL has begun integrating vector retrieval features, making it suitable for smaller scale applications.

These databases bridge the gap between structured data management and modern vector-based retrieval.

1.4.2.3 Modern Vector Databases

For small-scale or experimental use cases, vector search packages like FAISS (Facebook AI Similarity Search) and Annoy provide efficient similarity search and clustering. However, they lack the persistence, distributed computing, and advanced querying capabilities of full-fledged vector databases.

Vector databases are optimized for storing and querying high-dimensional embeddings, enabling semantic understanding in recommendation systems. Key options include:

1. **Pinecone**: A fully managed vector database offering real-time indexing, hybrid search, and metadata filtering.
2. **Weaviate**: An open-source vector database with built-in machine learning integrations and support for hybrid retrieval.
3. **Milvus**: A highly efficient open-source vector database designed for large-scale similarity search.

These systems excel in scenarios requiring low-latency, high-throughput retrieval of dense embeddings.

1.4.3 Retrieval in LLM-Powered Recommendation Systems

In LLM-powered recommendation systems, *retrieval* refers to the process of selecting a subset of relevant items from a large corpus to serve as candidate recommendations. This step is typically handled by a *candidate retrieval module*, which uses techniques such as dense vector similarity, sparse keyword matching, or hybrid methods to efficiently filter down the item pool. While traditional retrieval modules rely on pre-computed embeddings or interaction patterns, LLM-enhanced systems can incorporate richer signals—such as natural language queries, contextual information, or user profiles—into the retrieval process. By integrating LLMs into retrieval, these systems can better interpret user intent and dynamically adjust candidate selection before passing results to the ranking stage for final recommendation.

1.5 Encode-Decoder and Transformer Architecture

Encoder-decoder architectures and transformer models have become foundational in modern machine learning, originally excelling in tasks like summarization and translation. Recently, their ability to model complex input-output relationships and sequential data has made them increasingly valuable in recommendation systems. This section offers a brief introduction to these architectures as a foundation for their role in recommendation applications.

1.5.1 Encoder-Decoder Architecture

Encoder-decoder architectures are a class of neural networks designed to handle tasks that involve transforming one sequence into another (Sutskever et al., 2014).

- The *encoder* processes the input sequence (e.g., a user query or interaction history) and compresses it into a fixed-dimensional representation, often referred to as a context vector (Cho et al., 2014).

1.5 Encode-Decoder and Transformer Architecture

- The *decoder* then uses this representation to generate an output sequence (e.g., a recommendation or personalized summary).

Models like T5 (Raffel et al., 2020) and BART (Lewis et al., 2020) are prominent examples of encoder-decoder architectures, excelling in tasks such as text generation, translation, and summarization.

The framework of Encoder-Decoder architecture consists of two main components:

1. **Encoder Component**:
    - The encoder takes an input sequence of variable length and transforms it into a fixed-size *encoded representation* (often referred to as a "context vector" or "thought vector").
    - By processing each element of the input sequence (e.g., words or characters), the encoder captures contextual information to form a single vector that summarizes the input.

2. **Decoder Component**:
    - The decoder uses the encoded representation to generate the output sequence one element at a time.
    - It is *auto-regressive*, meaning it predicts each output element based on the encoded representation and previously generated elements.
    - This process continues until a special end-of-sequence token is generated or the maximum sequence length is reached.

Both the encoder and decoder often share similar architectures. In Fig. 1.4, for instance, an encoder processes the input text "Are you free tomorrow?" to produce a thought vector. The decoder then sequentially generates the response "Yes, what's up?" starting from a special <START> token and ending with <END>.

Table 1.3 summarizes three categories of encoder-decoder-based networks and their respective tasks:

- **Encoder-Only Models**
    - Suitable for understanding tasks, such as sentence classification and named entity recognition (NER).

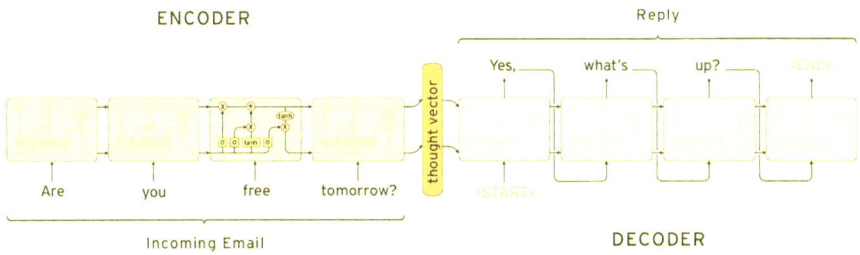

Fig. 1.4 Demonstration of encoder-decoder architecture in email reply example

Table 1.3 Three categories of encoder-decoder-based networks and their respective tasks

Model type	Examples	Tasks
Encoder-only	ALBERT, BERT, DistillBERT, RoBERTa	Topic classification, NER
Decoder-only	GPT, GPT-2, Transformer XL	Text generation
Encoder-Decoder	BART, T5	Machine translation, summarization, generative recommendation

- **Decoder-Only Models**
    - Ideal for generation tasks, like text completion or story generation.
    - Employ future masking, which prevents the model from using future tokens when predicting the next word. During training, teacher forcing ensures the model sees the complete text while masking future tokens to avoid information leakage.

- **Encoder-Decoder Models**
    - Designed for tasks where both *input and output* sequences are required, such as machine translation or summarization.

In the context of recommendation systems, encoder-decoder architectures are particularly well-suited for tasks that require generating personalized outputs, such as conversational recommendations, personalized content summaries, and multimodal recommendations (e.g., combining text and image data). Their ability to capture intricate relationships between input and output sequences makes them a powerful tool for enhancing user engagement and satisfaction.

1.5.2 Transformer Architecture

Transformers, introduced in the seminal work "Attention is All You Need" (Vaswani et al., 2017), revolutionized the field of machine learning by replacing traditional recurrent and convolutional layers with self-attention mechanisms. *Self-attention* allows the model to weigh the importance of different elements in a sequence relative to one another, enabling it to capture long-range dependencies and contextual relationships more effectively. This is particularly crucial in recommendation systems, where understanding user behavior, contextual signals, and item properties often involves processing long and complex input sequences.

The transformer architecture consists of stacked layers of self-attention and feedforward neural networks, making it highly scalable and parallelizable. These properties have led to the development of Large Language Models (LLMs) like GPT (Radford et al., 2018; Brown et al., 2020), BERT (Devlin et al., 2019), and their variants, which have achieved state-of-the-art performance across numerous natural language processing (NLP) tasks. In recommendation systems, transformers enable the modeling of user-item interactions, contextual information, and sequential patterns, resulting in highly personalized and context-aware recommendations.

1.5 Encode-Decoder and Transformer Architecture

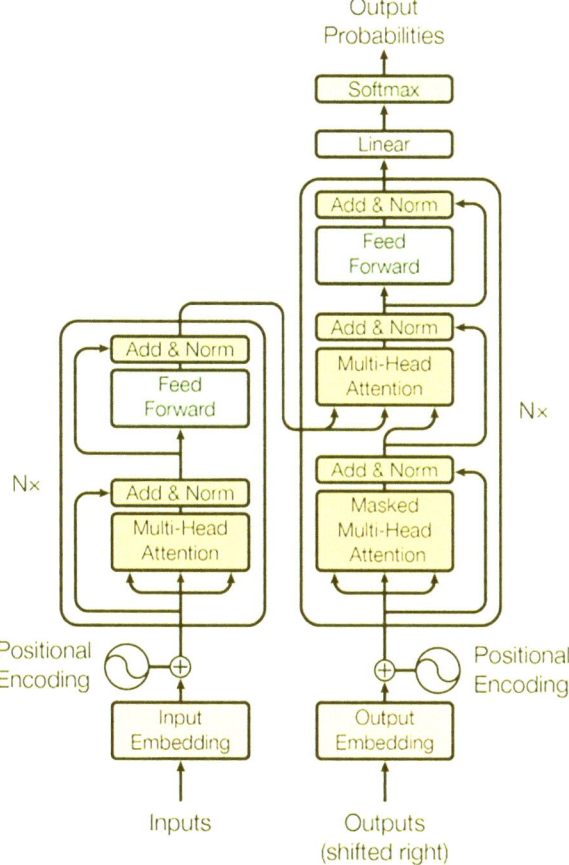

Fig. 1.5 Illustration of transformer architecture

1.5.2.1 Model Workflow

Figure 1.5 illustrates the transformer architecture workflow:

1. **Input Embedding**: Convert input tokens into dense vector representations.
2. **Positional Encoding**: Add positional information to embeddings to preserve token order.
3. **Encoder Stack**: Pass embeddings through multiple encoder layers, each containing:
   - Multi-head attention: Captures dependencies between tokens.
   - Feedforward network: Applies non-linear transformations.
   - Add and norm: Stabilizes training with residual connections and layer normalization.

4. **Output Embedding** (decoder): Convert output tokens (shifted right) into embeddings.
5. **Decoder Stack**: Process embeddings through multiple decoder layers, each containing:
   - Masked multi-head attention: Ensures predictions depend only on previous tokens.
   - Multi-head attention (encoder-decoder): Attends to encoder outputs.
   - Feedforward network: Applies non-linear transformations.
   - Add and norm: Stabilizes training.
6. **Linear Layer and Softmax**: Project decoder outputs into vocabulary-sized logits and computes probabilities for the next token.

1.5.2.2 Self-Attention and Q, K, V Mechanism

Self-attention computes a weighted sum of all tokens in a sequence, where the weights are determined by the relevance of each token to the others. This is achieved using three key vectors: Query (Q), Key (K), and Value (V).

1. **Query (Q)**: Represents the token for which we are computing attention. It is used to "query" other tokens for relevance.
2. **Key (K)**: Represents the tokens being compared to the query. It is used to compute the attention scores.
3. **Value (V)**: Represents the information carried by each token. It is used to compute the weighted sum.

Steps in Self-Attention

1. **Compute Attention Scores**: For each token, compute the dot product of its Query (Q) with the Keys (K) of all tokens. This measures the similarity between the query and each key

$$\text{Attention Score} = Q \cdot K^T$$

2. **Scale and Softmax**: Scale the scores by the square root of the dimension of the key vectors (to prevent large values) and apply the Softmax function to convert them into probabilities

$$\text{Scaled Attention Score} = \frac{Q \cdot K^T}{\sqrt{d_k}}$$

$$\text{Scaled Attention Weights} = \text{Softmax}\left(\text{Scaled Attention Score}\right)$$

3. **Weighted Sum**: Multiply the attention weights by the Values (*V*) to compute the final output for each token

$$\text{Output} = \text{Scaled Attention Weights} \cdot V$$

This mechanism allows the model to focus on the most relevant parts of the input sequence, capturing long-range dependencies and contextual relationships.

1.5.2.3 Positional Encoding

Positional encoding provides information about the order of tokens in a sequence, which is crucial since transformers lack recurrence or convolution. There are two main types of positional encoding: *absolute* and *relative*.

- **Absolute Positional Encoding**: In absolute positional encoding, a fixed set of sinusoidal functions is used to generate unique positional embeddings for each token position. These embeddings are added to the input embeddings. Absolute positional encoding is simple and effective but does not explicitly model relationships between positions.
- **Relative Positional Encoding**: Relative positional encoding focuses on the distance between tokens rather than their absolute positions. This allows the model to better capture relationships between tokens that are close to each other. Relative positional encoding is more flexible and often performs better in tasks where the relative order of tokens is more important than their absolute positions.

1.5.2.4 Categorization of Transformer Models

Transformer models can be classified based on their architecture and use cases:

- **GPT-Like Models (Auto-regressive)**: Examples include GPT, GPT-2, and GPT-3, primarily used for text generation and dialogue systems.
- **BERT-Like Models (Auto-encoding)**: Examples include BERT, RoBERTa, and ALBERT, used for sentence classification and extractive summarization.
- **BART/T5-Like Models (Seq2Seq)**: Examples include BART and T5, used for machine translation, summarization, and question answering.

    Table 1.4 contrasts GPT and BERT models across several dimensions.

Table 1.4 Contrast between GPT and BERT models in base architecture, learning paradigm, context, pre-training task, and applications

Dimension	GPT	BERT
Base architecture	Decoder-only transformer	Encoder-only transformer
Learning paradigm	Auto-regressive modeling	Masked language modeling (MLM)
Context	Uni-directional (left-to-right)	Bidirectional (context on both sides)
Pre-training task	Next-token prediction	Masked token prediction
Applications	Text generation, dialogue	Sentence understanding, QA

1.5.3 Transformers in LLM-Powered Recommendation Systems

The integration of encoder-decoder architectures and transformers into recommendation systems has opened up new possibilities for generative and sequential modeling. For instance:

- **Conversational Recommendations**: Encoder-decoder models can generate natural language responses in conversational interfaces, enabling interactive and dynamic recommendation experiences.
- **Personalized Summaries**: Transformers can analyze user interaction histories and generate personalized summaries or explanations for recommended items.
- **Multi-modal Recommendations**: By combining text, image, and other data modalities, these architectures can deliver richer and more diverse recommendations.

1.6 LLM Essentials

1.6.1 Scale and Core Capabilities

Large Language Models, such as GPT-3 and BERT, have revolutionized the field of natural language processing by demonstrating remarkable versatility and performance across a wide range of tasks. Their success is largely attributed to their scale, extensive training, and general applicability, which enable them to excel in diverse domains and applications.

- **Training Corpus**: LLMs are trained on extensive datasets comprising diverse sources such as books, articles, websites, and more. The scale can range from hundreds of gigabytes to several terabytes of text data.
- **Cost of Training**: Training LLMs is resource-intensive, often requiring significant computational power and time. This translates into substantial financial costs, sometimes reaching millions of dollars.
- **Number of Parameters**: The capabilities of LLMs are often tied to their size, measured in the number of parameters. Models can range from millions to hundreds of billions of parameters, with larger models generally exhibiting better performance and understanding.

1.6 LLM Essentials

LLMs are designed to be broadly applicable across various domains and tasks, leveraging their extensive training and robust architecture. Their ability to understand and generate human-like text makes them suitable for a wide range of applications in natural language processing, content creation, customer service, and beyond. Some key applications include:

- **Sentiment Analysis**: LLMs can analyze text to determine the sentiment expressed, identifying whether the opinion is positive, negative, or neutral. This capability is valuable for applications such as market research, social media monitoring, and customer feedback analysis.
- **Named Entity Recognition**: LLMs excel at identifying and classifying proper nouns and specific entities within text, such as names of people, organizations, and locations. This is crucial for tasks like information extraction, data categorization, and knowledge graph construction.
- **Dialogue Systems**: LLMs are adept at generating coherent and contextually relevant responses in conversational settings, making them ideal for chatbots and virtual assistants.
- **Translation**: They can translate text between multiple languages, preserving the meaning and nuance of the original content.
- **Summarization**: LLMs can generate concise summaries of longer texts, capturing key points and essential information efficiently.

1.6.2 Emergent Abilities

As Large Language Models continue to scale, they exhibit *emergent abilities*—unexpected capabilities that arise as the model size increases (Wei et al., 2022a). These abilities enable LLMs to perform a wide range of tasks in a flexible and efficient manner, making them highly effective for LLM-powered recommendation systems. Here, we explore three key emergent abilities: In-Context Learning (ICL) (Brown et al., 2020), Instruction Following (Ouyang et al., 2022), and Chain-of-Thought (CoT) Reasoning (Wei et al., 2022b).

Figure 1.6 demonstrates how these emergent abilities become more pronounced as model size increases, particularly around the 100 billion parameter mark. This finding underscores the importance of scaling in developing more capable and versatile AI systems.

1.6.2.1 In-Context Learning

In-Context Learning (ICL) enables LLMs to adapt and perform tasks based on context provided within the prompt, without the need for task-specific training data. This allows LLMs to learn from the examples provided within the prompt, making them adaptable to new tasks with minimal input. Two prominent examples of in-context learning are *zero-shot learning* and *few-shot learning,* which are particularly relevant in the context of recommendation systems.

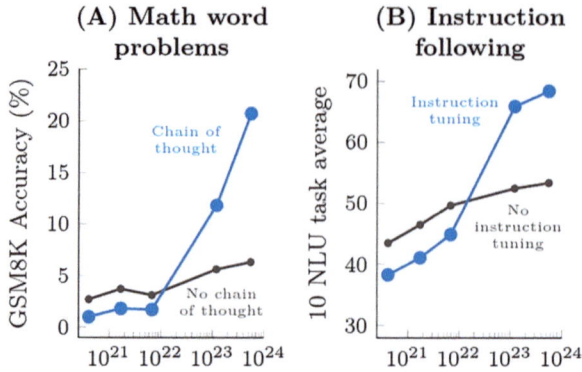

Fig. 1.6 Demonstration of how (**a**) CoT and (**b**) instruction-following abilities improve as model size increases

- **Zero-Shot Learning**: In zero-shot learning, the model performs a task without any explicit examples, relying solely on its pre-trained knowledge and the instructions provided in the prompt. Here is an example in recommendation context:

```
Prompt: "Recommend laptops under $1000 with a battery life
of over 8 hours."
Model Output: "Here are some laptops that meet your
criteria: Acer Aspire 5,
Lenovo Ideapad 3, and Dell Inspiron 14."
```

- **Few-Shot Learning**: By providing a few examples, *few-shot learning* helps the model better understand the task and improve its output. Here is an example of few-shot learning in recommendation context:

```
Prompt: "I like budget-friendly laptops for students with
long battery life.
Here's an example: 'Acer Aspire 5, $500, 10-hour battery.'
Recommend similar laptops."
Model Output: "Here are some similar laptops: HP Pavilion
x360, $450, 9-hour
battery; Lenovo ThinkPad, $600, 12-hour battery."
```

In-context learning enables LLMs to generate relevant recommendations based on minimal context, improving the flexibility of the recommendation system.

1.6.2.2 Instruction Following

LLMs can *understand and execute tasks* given to them via natural language instructions. This ability allows them to perform tasks without explicit retraining, simply by following the instructions provided.

Here's how this applies, using recommendation examples:

LLMs can follow direct instructions for specific outputs:

- **Prompt**: "Provide me with a list of top-rated budget smartphones for gaming."
- **Model Output**: "Here are some top-rated gaming smartphones within budget: Xiaomi Redmi Note 10, Realme Narzo 30 Pro, and Poco X3 Pro."

They can also adapt to tasks defined by specific criteria within the instruction, demonstrating flexibility:

- **Prompt**: "Generate a recommendation list for users who prefer lightweight laptops with good battery life."
- **Model Output**: "Based on user preferences, I recommend: Apple MacBook Air, Dell XPS 13, and HP Spectre x360."

Essentially, the ability to follow instructions allows LLMs to easily generate recommendations (and perform many other tasks) tailored to diverse needs, showcasing their adaptability through natural language understanding.

1.6.2.3 Chain-of-Thought (CoT) Reasoning

Chain-of-Thought (CoT) Reasoning helps LLMs break down complex tasks into smaller, more manageable steps, improving their ability to solve intricate problems and enhancing the interpretability of their outputs. This ability is especially useful in recommendation systems where multiple factors (e.g., price, features, ratings) need to be considered when generating personalized suggestions.

- **Problem Solving with CoT**: CoT reasoning guides the model to articulate intermediate steps, making its reasoning process more transparent and structured.
- **Example in Recommendation Systems**:
  - **Prompt**: "Recommend a laptop for a student who needs a lightweight laptop with good performance and a budget under $800."
  - **CoT Reasoning**:

    Step 1: **Filter laptops under $800**: Acer Aspire 5, Lenovo IdeaPad, HP Pavilion.
    Step 2: **Check weight**: Ensure the laptop is under 4 pounds.
    Step 3: **Check performance**: Ensure the laptop has at least an Intel Core i5 processor.
    Model Output: "Based on these criteria, I recommend the Acer Aspire 5, Lenovo IdeaPad 3, and HP Pavilion 14."

CoT reasoning helps the model break down the decision-making process, improving the quality and relevance of the generated recommendations.

1.7 LLM Pre-training, Post-training, and Inference

Figure 1.7 illustrates the standard training pipeline for large language models (LLMs):

- **Pre-training**: A base model is first trained on massive unlabeled text corpora using self-supervised learning.
- **Supervised Fine-Tuning (SFT)**: The base model is further adapted to specific tasks using annotated datasets.
- **RLHF (Reinforcement Learning from Human Feedback)**: The fine-tuned model is aligned with human preferences using reinforcement learning guided by a reward model.

This multi-stage process enables LLMs to be both linguistically fluent and task-aligned.

1.7.1 Pre-training

Pre-training is the initial phase of training an LLM, where the model learns the fundamentals of human language by processing large-scale text datasets. These datasets often include diverse sources such as Wikipedia, news articles, books, and web content. During pre-training, the model is trained to predict the next word in a sentence (in the case of auto-regressive models like GPT) or to fill in missing words (in the case of masked language models like BERT). This process allows the model to develop a general understanding of grammar, syntax, semantics, and world knowledge.

Pre-training involves training the model on massive amounts of text data using unsupervised or self-supervised learning objectives. For example:

- In *auto-regressive models*, the model predicts the next token in a sequence given the previous tokens.
- In *masked language models*, the model predicts masked tokens within a sequence based on the surrounding context.

Pre-training equips the model with a foundational understanding of language, enabling it to generalize across a wide range of tasks. Without pre-training, the model would lack the linguistic and contextual knowledge necessary to perform effectively on downstream tasks.

1.7 LLM Pre-training, Post-training, and Inference 25

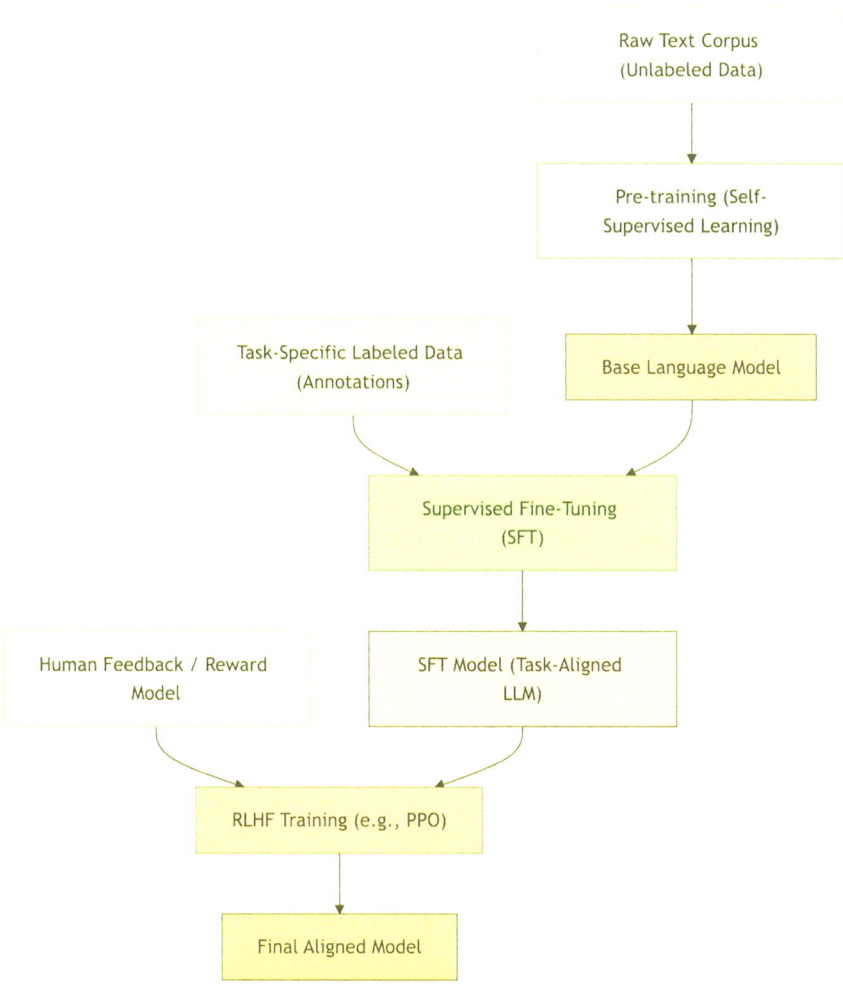

Fig. 1.7 Pre-training and post-training LLMs

1.7.2 Supervised Fine-Tuning

Supervised fine-tuning (SFT) is a critical step in adapting pre-trained large language models (LLMs) to specific tasks, domains, or skills. While pre-trained models like GPT-4 or LLaMA possess broad linguistic capabilities, SFT tailors these models to excel in targeted applications. Key motivations include:

1. **Task Adaptation**: SFT enables LLMs to specialize in specific tasks, such as recommendation systems, where the model is fine-tuned on labeled user-item interaction data to predict preferences. This ensures the model aligns with task-specific objectives, like generating personalized recommendations.

2. **Domain Knowledge Integration**: SFT instills domain-specific knowledge, such as medical terminology for healthcare applications or legal jargon for contract analysis. For example, fine-tuning GPT-4 on medical literature improves its ability to answer patient queries accurately.
3. **Skill Enhancement**: SFT can improve specific skills like conversational reasoning, logical inference, or multi-turn dialogue. For instance, fine-tuning on conversational datasets enhances the model's ability to maintain context and provide coherent responses.
4. **Behavioral Alignment**: SFT aligns the model's outputs with desired behaviors, such as generating safe, ethical, or user-friendly responses. This is particularly important for applications like customer support or educational tutoring.

1.7.2.1 SFT Workflow

As illustrated in Fig. 1.8, the SFT process involves the following steps:

1. **Select a Pre-trained Model**. Choose a model based on task requirements, size, and pre-training data:
   - **Task Requirements**. For example, for tasks requiring deep semantic understanding like user reviews or product descriptions, we can choose models like

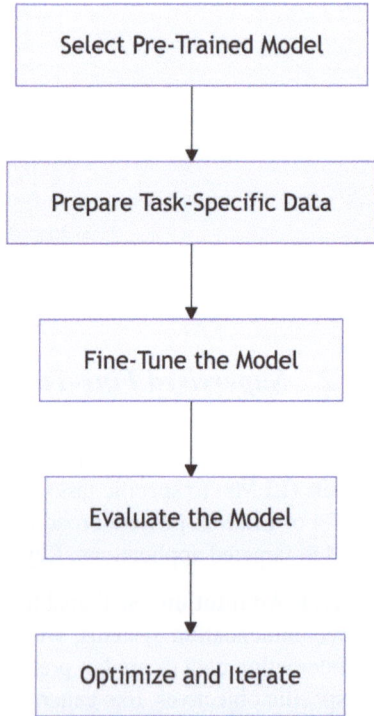

Fig. 1.8 Workflow for supervised fine-tuning

BERT or RoBERTa, while for tasks requiring text generation like rewriting queries or personalized emails, we can use GPT-based models.
- **Model Size**: In cases when we prioritize accuracy over latency and computation cost, we can choose large based models, while if low latency or cost are strong concerns, small models like DistillBERT or TinyBERT are preferred.
- **Pre-training Data**: Ensure the base model is pre-trained on data relevant to your domain (e.g., e-commerce, social media). For example, in an e-commerce setting, we can use a model pre-trained on product reviews.

2. **Prepare Task-Specific Data**:
   - Use domain-specific data like product catalogs, reviews, or interaction logs.
   - Ensure data cleanliness and label quality.

3. **Fine-Tune the Model**:
   - Adjust hyperparameters (e.g., learning rate, batch size).
   - Freeze early layers to retain general knowledge or fine-tune all layers for task-specific adaptation.

4. **Evaluate and Optimize**:
   - Test the model on a validation or test set.
   - Use metrics like precision, recall, or NDCG to evaluate performance.
   - Iteratively optimize hyperparameters and data preprocessing.

1.7.2.2 Existing Frameworks for SFT

Several frameworks and algorithms simplify the SFT process:

1. **Hugging Face Transformers**:
   - Provides pre-trained models (e.g., BERT, GPT) and tools for fine-tuning.
   - Example:

```
from transformers import AutoModelForSequenceClassification,
Trainer, TrainingArguments
model = AutoModelForSequenceClassification.
from_pretrained("bert-base-uncased")
training_args = TrainingArguments(output_dir="./results",
learning_rate=2e-5, per_device_train_batch_size=16)
trainer = Trainer(model=model,args=training_args,
train_dataset=train_dataset)
trainer.train()
```

2. **PyTorch Lightning**:
   - Simplifies training loops and supports distributed training.
   - Example:

```
from pytorch_lightning import Trainer
trainer = Trainer(max_epochs=3, gpus=1)
trainer.fit(model, train_dataloader)
```

3. **LoRA (Low-Rank Adaptation)**:
   - Fine-tunes only a small subset of parameters, reducing computational cost.
   - Example: Use LoRA to adapt GPT-3 for recommendation tasks without retraining the entire model.

4. **Parameter-Efficient Fine-Tuning (PEFT)**:
   - Techniques like adapters or prefix tuning reduce the number of trainable parameters.
   - Example: Use adapters to fine-tune T5 for query rewriting in recommendation systems.

Supervised fine-tuning (SFT), while essential for task adaptation, faces the following challenges:

- **Objective Misalignment**: The goals of pre-training and SFT often conflict. For example, GPT-style models are pre-trained for next-token prediction, whereas SFT optimizes for task-specific losses (e.g., instruction-response alignment). This mismatch risks eroding the model's general linguistic capabilities while prioritizing narrow task performance.
- **Overfitting Risk**: Fine-tuning datasets are inherently smaller than pre-training corpora, increasing susceptibility to overfitting. Without careful hyperparameter tuning (e.g., reduced learning rates) or regularization, the model may lose its ability to generalize, particularly in dynamic domains like recommendation systems where user preferences evolve rapidly.
- **Difficulty Incorporating Preferential Feedback**: SFT struggles to incorporate nuanced user preferences or subjective feedback, such as ranking multiple responses by quality. This limitation motivates the use of reinforcement learning with human feedback (RLHF), which refines models based on iterative human evaluations.
- **Data Scarcity and Cost**: High-quality labeled datasets for SFT are often scarce or expensive to create, particularly for niche domains or specialized tasks.

1.7.3 Reinforcement Learning with Human Feedback

Reinforcement Learning with Human Feedback (RLHF) is a post-training technique that uses reinforcement learning to align the model's outputs with human preferences. It is particularly useful for tasks where the desired behavior is complex or subjective, such as generating conversational responses or recommendations that align with user satisfaction. RLHF is usually carried out in the following steps:

- Human annotators provide *feedback* on the model's outputs, ranking them based on quality, relevance, or alignment with desired behavior.
- A *reward model* is trained to predict these human preferences.
- The LLM is then *fine-tuned using reinforcement learning*, where it learns to maximize the reward predicted by the reward model.

RLHF ensures that the model's outputs are not only accurate but also aligned with human values and preferences. This step is essential for improving user satisfaction and trust, especially in applications like conversational recommendations or personalized content generation.

1.7.4 LLM Inference

Inference is the phase where trained large language models (LLMs) are used to generate predictions based on user input. Whether deploying locally or calling an API, the process begins with loading the model and tokenizer. Here are two typical approaches:

Loading a Local Checkpoint (e.g., Hugging Face Transformers)

```
from transformers import AutoTokenizer, AutoModelForCausalLM
tokenizer = AutoTokenizer.from_pretrained("gpt2")
model = AutoModelForCausalLM.from_pretrained("gpt2")
```

Using an LLM API (e.g., OpenAI)

```
import openai
openai.api_key = "your-api-key"
response = openai.ChatCompletion.create(
    model="gpt-4",
    messages=[{"role": "user", "content": "Recommend a sci-fi book for teens"}]
)
print(response["choices"][0]["message"]["content"])
```

Next, we explore key techniques and optimizations that enable efficient and scalable inference, including auto-regressive and speculative decoding, architecture-specific strategies, and optimization methods like batching, caching, and quantization.

1.7.4.1 Auto-Regressive and Speculative Decoding

Auto-regressive decoding, used in models like GPT, generates text token-by-token, predicting the next word based on previous words. While effective, this sequential process can result in high latency. Speculative decoding addresses this by using a smaller "draft" model to predict multiple tokens ahead, which are then verified by the larger model. This optimization reduces inference latency by minimizing the number of forward passes, making it ideal for real-time applications. We will briefly touch on speculative decoding in our tutorial in Sect. 1.8.

1.7.4.2 Architecture-Specific Inference

Different architectures optimize inference in unique ways. GPT models excel at auto-regressive decoding with techniques like top-k and nucleus sampling, ensuring high-quality text generation. Llama focuses on efficiency with dynamic batching and sparse attention, reducing computational overhead. Mixture of Experts (MoE) models use specialized sub-models (experts) activated only for relevant inputs, enabling scalable and efficient inference. Each architecture offers distinct trade-offs, catering to specific use cases and deployment scenarios.

1.7.4.3 Batching and Caching

Batching processes multiple inputs simultaneously, improving throughput by leveraging parallel computation. Caching reuses intermediate computations to avoid redundant calculations, further enhancing efficiency. Together, these techniques optimize inference for real-time applications, ensuring faster and more scalable deployments. We will cover more details on caching in Sect. 4.3.2.

1.7.4.4 Quantization and Model Compression

Quantization reduces model size and computational requirements by lowering the precision of weights (e.g., from 32-bit to 8-bit). Model compression techniques like pruning remove redundant weights, further shrinking the model. These methods enable deployment on resource-constrained devices, making LLMs more accessible for edge and mobile applications. We will cover more details on quantization and model compression in Sect. 4.3.2.

1.8 Tutorial: Understanding Tokenization and Transformer Model

1.8.1 Overview

This tutorial uses a toy example to explain how transformer models process text, from tokenization to advanced inference methods. It covers key concepts such as attention mechanisms, hidden states, and auto-regressive decoding.

Goal of This Tutorial

1. Understand the role of *tokenization* in preparing text data for transformer models.
2. Explore how *transformer models* process tokenized inputs using attention mechanisms.
3. Visualize *attention weights* to observe how transformers focus on specific tokens.
4. Examine *inference techniques* like auto-regressive decoding and discuss advanced optimization methods.

We show a condensed version of this tutorial in the book text. The full version of the code is available at: https://github.com/qqwjq1981/springer-LLM-recommendation-system

1.8.2 Experimental Design

The tutorial is structured into four steps, each focusing on a specific aspect of tokenization and transformer models:

Step 1: Tokenization Basics
In this step, we start with a simple sentence, "Transformers have revolutionized natural language processing!", and use a pre-trained tokenizer to break text into tokens.

Step 2: Understanding Transformers
We load a pre-trained transformer model (e.g., BERT) and pass tokenized input ids through the model.

Step 3: Visualizing Attention Weights
We enable attention outputs in the transformer model, and visualize attention weights using heatmaps to observe how the model focuses on specific tokens.

Step 4: Auto-regressive Decoding
Generate text using GPT-2 and examine the top-k decoded sequences to demonstrate the diversity of outputs.

We used our customized implementation of auto-regressive and speculative decoding.

1.8.3 Results and Analysis

Step 1: Tokenization Basics

```
Original Text: Transformers have revolutionized natural
language processing!
Tokens: ['transformers', 'have', 'revolution', '##ized',
'natural', 'language', 'processing', '!']
Input IDs: tensor([[  101, 19081,  2031,  4329,  3550,
         3019,  2653,  6364,   999,   102]])
```

The tokenizer successfully breaks the text into meaningful subword units, and each token is converted to a numeric ID for the model to process.

Step 2: Understanding Transformers

```
# Extract the hidden states
hidden_states = outputs.last_hidden_state
print("Shape of Hidden States:", hidden_states.shape)
#Output
# Shape of Hidden States: torch.Size([1, 10, 768])
```

hidden_states is the output of the last layer of the BERT model. It represents contextual embeddings (or vector representations) of each token in your input, after processing through all BERT layers. The shape of hidden_states is [batch_size, sequence_length, hidden_size].

- batch_size: Number of samples (sentences) you fed into the model at once.
- sequence_length: Number of tokens in each input sentence (after tokenization).
- hidden_size: Size of the vector representation for each token (for BERT-base, it is *768*).

Step 3: Visualizing Attention Weights

Figure 1.9 shows the attention weights heatmap.

Query tokens represent the tokens being generated, and key tokens represent the tokens in the context. Suppose you're generating a sentence:

> The cat sat on the ...

Now the model needs to generate the *next word*, like "mat".

- The *query* is the *position where the next token will go*.
- The *keys* (and *values*) come from the embeddings of the *previous tokens*: "The", "cat", "sat", "on", "the".

1.8 Tutorial: Understanding Tokenization and Transformer Model   33

The *attention Heatmap* shows higher weights for semantically related tokens (e.g., "transformers" and "natural language processing").

Step 4: Auto-regressive decoding

```
input_text = "The winter in California is "
Top-k decoded sequences:
1: The winter in California is iced. It's hot. It's cold.
It's cold.
2: The winter in California is iced up, and it's going to be
colder than it used to
3: The winter in California is iced and cold, and there is
no water to drink.
4: The winter in California is iced, and it's hard to find a
place that doesn't have
5: The winter in California is iced, and it's cold enough
that I can't stand it.
```

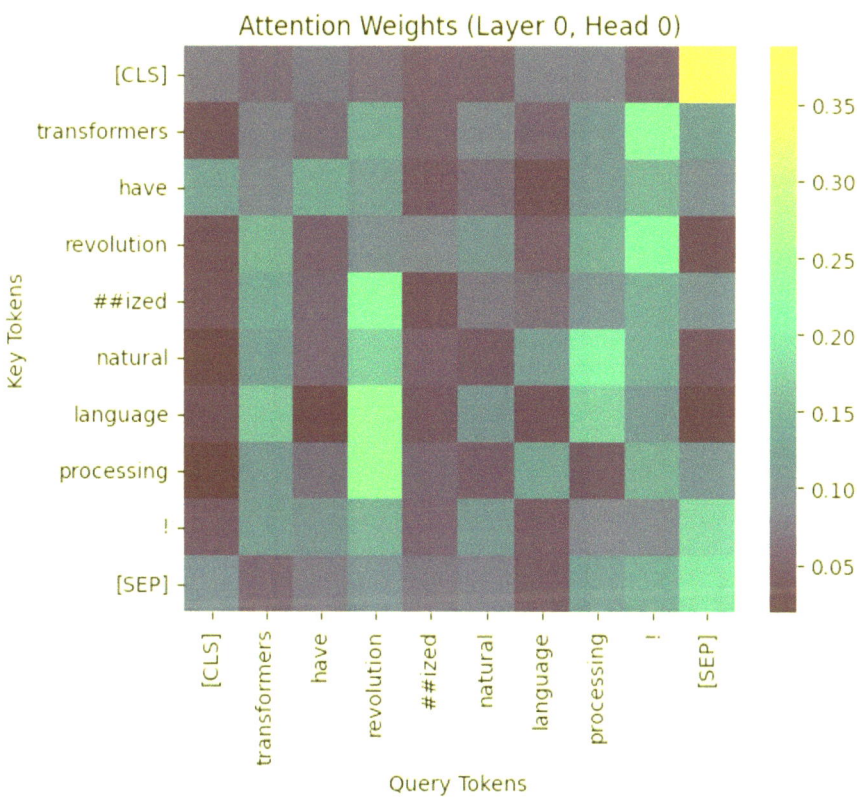

Fig. 1.9 Attention weight heatmap

In auto-regressive decoding, we demonstrate that GPT-style models generate entire sequences by predicting one token at a time. The above output shows the top-*k* decoded sequences, they share similarity in stating the winter in California is iced, but with variations from one to another.

1.8.3.1 Advanced Methods

While the above steps introduce key concepts, it's also helpful to briefly highlight a couple of advanced methods that are shaping the next generation of recommendation and generation systems:

Custom Tokenizers

Sometimes, pre-trained tokenizers aren't enough:

- In *domain-specific applications* (e.g., legal, medical), you may want to better capture rare terms.
- *Multilingual settings* benefit from tokenizers trained on diverse languages.
- You might also train *smaller, efficient tokenizers* tailored for faster deployment.

Custom tokenizers can be created using libraries like Hugging Face's tokenizers, trained directly on your own text corpus.

Speculative Decoding

This method is used to *speed up text generation* without fully relying on large models:

- A smaller *drafter* model generates candidate tokens quickly.
- A larger *target* model *verifies or corrects* those tokens in parallel.
- It's a clever way to reduce latency while maintaining output quality.

Although not covered in detail in this tutorial, speculative decoding is a powerful optimization strategy in real-world applications.

1.8.4 Conclusion

This tutorial provided a foundational walkthrough of tokenization and auto-regressive decoding in transformer models. We explored how BERT tokenizers process text and how GPT-style models generate sequences step by step. Along the way, we visualized attention and token probabilities to better understand model behavior. Finally, we briefly highlighted advanced techniques like custom tokenizers and speculative decoding, offering a glimpse into more efficient and domain-adaptive applications.

1.9 Second Tutorial: Understanding Content Embedding and Retrieval

1.9.1 Overview

Retrieving relevant news articles from large datasets is a critical task for applications like personalized news feeds and content recommendation systems. This tutorial demonstrates how to generate embeddings for news article summaries, visualize them using t-SNE, and evaluate the performance of sparse, dense, and hybrid retrieval methods.

Goals of This Tutorial

1. Understand the role of embeddings in capturing semantic relationships between content.
2. Generate embeddings for textual data using pre-trained LLMs.
3. Perform content retrieval using Approximate Nearest Neighbor (ANN) search with these embeddings.

This example builds on the theoretical foundations of embeddings, retrieval systems, and evaluation metrics discussed in earlier chapters, providing a practical implementation for real-world applications. We show a condensed version of this tutorial in the book text. The full version of the code is available at: https://github.com/qqwjq1981/springer-LLM-recommendation-system

1.9.2 Experimental Design

The study is designed to evaluate the effectiveness of sparse, dense, and hybrid retrieval methods for news article retrieval.

1. **Data Preparation**:
   - For this project, we used the Kaggle BBC News Article Summary Dataset, which consists of news articles labeled into five distinct categories: ["business", "tech", "entertainment", "sport", "politics"].
   - Embeddings are generated using a sentence-transformer model (all-MiniLM-L6-v2).
2. **Visualization**:
   - t-SNE is applied to reduce the dimensionality of embeddings to 2D for visualization.
   - Points in the t-SNE plot are colored by the primary topic of the news article.

3. **Retrieval Methods**:
   - **Sparse Retrieval**: Use BM25 to retrieve articles based on keyword matching.
   - **Dense Retrieval**: Use cosine similarity on embeddings to retrieve semantically similar articles.
   - **Hybrid Retrieval**: Combine BM25 and dense retrieval scores using a weighted sum.
4. **Evaluation Framework**:
   - Precision@K and Recall@K are computed for each retrieval method.
   - Retrieval time is measured to assess efficiency.

1.9.3 Results and Analysis

1.9.3.1 t-SNE Visualization

The t-SNE plot in Fig. 1.10 shows clear clustering of articles by primary topic, indicating the article clusters are nearly separable in the embeddings space.

1.9.3.2 Results Table

The evaluation metrics are presented in Table 1.5, categorized by the primary article topic.

1. **Precision**:
   - BM25 performs well for business, entertainment, and politics at $k = 10$, but its precision drops for tech and sport.
   - *Dense* models excel for tech but struggle with entertainment and sport.
   - *Hybrid* models consistently perform well across most categories, especially for tech and politics.
2. **NDCG**:
   - BM25 achieves high NDCG scores for tech and politics, indicating strong ranking quality.
   - Dense models perform well for tech but poorly for entertainment.
   - Hybrid models show balanced performance, with high NDCG scores for tech, politics, and business.
3. **Category-Specific Trends**:
   - **Tech**: All models perform well, with Dense and Hybrid achieving perfect NDCG at $k = 10$.
   - **Entertainment**: BM25 outperforms Dense and Hybrid, especially at $k = 10$.
   - **Politics**: BM25 and Hybrid achieve perfect NDCG at $k = 10$, while Dense lags slightly.

1.9 Second Tutorial: Understanding Content Embedding and Retrieval

Fig. 1.10 t-SNE visualization of article embeddings, colored by primary topic category

Table 1.5 Precision and NDCG @10 for three retrieval approaches: BM25, dense, and hybrid retrieval

Category	Model	Precision@10	NDCG@10	Precision@20	NDCG@20
Business	BM25	0.9	0.9667	0.9	0.9679
	Dense	0.4	0.8036	0.35	0.7743
	Hybrid	0.8	0.8632	0.8	0.8802
Tech	BM25	0.3	1.0	0.15	1.0
	Dense	1.0	1.0	1.0	1.0
	Hybrid	1.0	1.0	0.95	0.9967
Entertainment	BM25	0.8	0.9963	0.5	0.9761
	Dense	0.0	0.0	0.05	0.2560
	Hybrid	0.2	0.4228	0.25	0.4884
Sport	BM25	0.5	0.7135	0.3	0.7183
	Dense	0.3	0.5724	0.5	0.6460
	Hybrid	0.5	0.6092	0.4	0.6476
Politics	BM25	1.0	1.0	0.75	0.9991
	Dense	0.8	0.7985	0.65	0.8217
	Hybrid	1.0	1.0	0.9	0.9929

1.9.4 Conclusion

This tutorial illustrates the effectiveness of embedding-based and hybrid retrieval methods for the task of news article retrieval. The key insights are as follows:

- **Semantic Embeddings**: Embedding representations effectively capture latent semantic relationships between articles, as evidenced by structured clustering in the t-SNE visualization.
- **Hybrid Retrieval**: Combining sparse (e.g., BM25) and dense (e.g., embedding-based) retrieval techniques yields the most favorable trade-off among precision, recall, and response time.
- **Scalability and Deployment**: The proposed retrieval pipeline demonstrates strong scalability characteristics, making it suitable for deployment in real-time or large-scale retrieval environments.

References

Bahdanau, D., Cho, K., & Bengio, Y. (2014). Neural machine translation by jointly learning to align and translate. *ArXiv:1409.0473*.

Baker, J. K. (1975). Stochastic modeling for automatic speech understanding. In D. R. Reddy, ed., Speech Recognition. Academic Press.

Brown, T., Mann, B., Ryder, N., Subbiah, M., Kaplan, J., Dhariwal, P., et al. (2020). Language models are few-shot learners. In *Advances in neural information processing systems (NeurIPS)* (Vol. 33, pp. 1877–1901).

Cer, D., Yang, Y., Kong, S., Hua, N., Limtiaco, N., John, R. S., et al. (2018). *Universal sentence encoder*. arXiv preprint arXiv:1803.11175.

Cho, K., van Merriënboer, B., Gulcehre, C., Bahdanau, D., Bougares, F., Schwenk, H., & Bengio, Y. (2014). Learning phrase representations using RNN encoder-decoder for statistical machine translation. In *EMNLP*.

Chomsky, N. (1957). *Syntactic structures*. Mouton & Co.

Conneau, A., Kiela, D., Schwenk, H., Barrault, L., & Bordes, A. (2017). Supervised learning of universal sentence representations from natural language inference data. In *EMNLP 2017* (pp. 670–680).

Devlin, J., Chang, M.-W., Lee, K., & Toutanova, K. (2018). BERT: Pre-training of deep bidirectional transformers for language understanding. *ArXiv:1810.04805*.

Devlin, J., Chang, M.-W., Lee, K., & Toutanova, K. (2019). BERT: Pre-training of deep bidirectional transformers for language understanding. In *Proceedings of NAACL-HLT* (pp. 4171–4186).

Elman, J. L. (1990). Finding structure in time. *Cognitive Science, 14*(2), 179–211.

Hochreiter, S., & Schmidhuber, J. (1997). Long short-term memory. *Neural Computation, 9*(8), 1735–1780.

Jelinek, F. (1976). Continuous speech recognition by statistical methods. *Proceedings of the IEEE, 64*(4), 532–556.

Kim, Y. (2014). Convolutional neural networks for sentence classification. In *Proceedings of the 2014 Conference on Empirical Methods in Natural Language Processing (EMNLP)* (pp. 1746–1751). Association for Computational Linguistics.

Kim, Y., Jernite, Y., Sontag, D., & Rush, A. M. (2016). Character-aware neural language models. In *AAAI*.

References

Kudo, T., & Richardson, J. (2018). SentencePiece: A simple and language independent subword tokenizer and detokenizer for neural text processing. In *EMNLP*.

Lewis, M., Liu, Y., Goyal, N., Ghazvininejad, M., Mohamed, A., Levy, O., et al. (2020). BART: Denoising sequence-to-sequence pre-training for natural language generation, translation, and comprehension. In *ACL*.

Markov, A. A. (1913). Essai d'une recherche statistique sur le texte du roman "Eugene Onegin" illustrant la liaison des epreuve en chain ('Example of a statistical investigation of the text of "Eugene Onegin" illustrating the dependence between samples in chain'). Izvistia Impcratorskoi Akademii Nauk (Bulletin de l'Academie Imp ´ eriale´ des Sciences de St.-Petersbourg)´, 7:153–162.

Mikolov, T., Chen, K., Corrado, G., & Dean, J. (2013). *Efficient estimation of word representations in vector space*. arXiv preprint arXiv:1301.3781.

Ouyang, L., Wu, J., Jiang, X., Almeida, D., Wainwright, C., Mishkin, P., et al. (2022). Training language models to follow instructions with human feedback. In *NeurIPS*.

Pennington, J., Socher, R., & Manning, C. D. (2014). GloVe: Global vectors for word representation. In *EMNLP* (pp. 1532–1543).

Peters, M. E., Neumann, M., Iyyer, M., Gardner, M., Clark, C., Lee, K., & Zettlemoyer, L. (2018). Deep contextualized word representations. In *NAACL 2018* (pp. 2227–2237).

Radford, A., Narasimhan, K., Salimans, T., & Sutskever, I. (2018). Improving language understanding by generative pre-training. In *OpenAI*.

Raffel, C., Shazeer, N., Roberts, A., Lee, K., Narang, S., Matena, M., et al. (2020). Exploring the limits of transfer learning with a unified text-to-text transformer. *Journal of Machine Learning Research, 21*(140), 1–67.

Reimers, N., & Gurevych, I. (2019). Sentence-BERT: Sentence embeddings using Siamese BERT-networks. In *EMNLP*.

Schuster, M., & Nakajima, K. (2012). Japanese and Korean voice search. In *IEEE ICASSP*.

Sennrich, R., Haddow, B., & Birch, A. (2016). Neural machine translation of rare words with subword units. In *ACL*.

Sutskever, I., Vinyals, O., & Le, Q. V. (2014). Sequence to sequence learning with neural networks. In *NeurIPS* (Vol. 27, pp. 3104–3112).

Vaswani, A., Shazeer, N., Parmar, N., Uszkoreit, J., Jones, L., Gomez, A. N., Kaiser, Ł., & Polosukhin, I. (2017). Attention is all you need. In *Advances in neural information processing systems (NeurIPS)* (Vol. 30, pp. 5998–6008).

Wei, J., Wang, X., Schuurmans, D., Bosma, M., Ichter, B., Xia, F., et al. (2022a). *Emergent abilities of large language models*. Transactions on Machine Learning Research.

Wei, J., Wang, X., Schuurmans, D., Bosma, M., Xia, F., Chi, E., et al. (2022b). Chain of thought prompting elicits reasoning in large language models. In *NeurIPS*.

Weizenbaum, J. (1966). ELIZA—A computer program for the study of natural language communication between man and machine. *Communications of the ACM, 9*(1), 36–45.

Chapter 2
From Traditional to LLM-Powered Recommendation Systems

This chapter explores the transition from traditional recommendation systems to approaches powered by large language models (LLMs). It begins with an overview of conventional pipelines—covering content understanding, user modeling techniques such as collaborative filtering and matrix factorization, candidate retrieval, and ranking strategies. The chapter then highlights key limitations of traditional methods and explains how LLMs address these challenges through unified representations, enhanced personalization, and improved scalability. Various LLM-based paradigms are examined, including models that augment existing systems as well as those that serve as end-to-end recommenders. A hands-on tutorial using the MovieLens dataset illustrates this evolution in practice, comparing traditional and LLM-based methods through empirical results.

2.1 Recommendation System Workflow

Since their introduction in the late 1990s, recommendation systems have become essential components in modern computational frameworks for information retrieval and personalized services. These systems were initially based on basic collaborative filtering methods and used user-item interaction matrices to suggest similar items. Over time, recommendation systems have evolved significantly alongside the growth of data, algorithmic innovation, and computational power. The rise of deep learning models and graph-based approaches have allowed for innovations in modeling user preferences, item features, and contextual factors, addressing key challenges such as recommendation quality, diversity, and interpretability.

Today, recommendation systems are foundational technologies across a wide range of digital platforms, including e-commerce, social media, and streaming services.

- In *commercial applications*, they help personalize product discovery by integrating multiple signals such as purchase history, session behavior, and user intent across domains.
- In *content platforms*, neural recommendation models use techniques like attention mechanisms and temporal modeling to create hyper-personalized feeds, enhancing user retention, engagement, and long-term satisfaction.

The use of artificial intelligence, especially transformer-based models and reinforcement learning, has further advanced real-time recommendations, multi-objective optimization, and explainable systems.

Figure 2.1 illustrates a typical recommendation system workflow. The process begins by collecting data from two main sources: the *item corpus* and *user history and context*. These inputs feed into core system modules: *content understanding* analyzes items, while *user modeling* captures user preferences. The system then performs *candidate retrieval* to filter millions of items down to a few hundred, which are further refined through *ranking*. The top-ranked items are presented as recommendations, and the *evaluation and feedback loop* continuously updates the system based on user interactions to improve future recommendations.

Next, we provide an in-depth examination of each component, discussing their roles and methodologies in creating effective personalized recommendations.

2.1.1 Content Understanding

Content understanding is a key step in recommendation systems, involving the extraction, interpretation, and structuring of item features to improve recommendation relevance. This process incorporates several common techniques:

- **Item Identification**: Assigning unique identifiers to each item to be recommended.
- **Metadata Extraction**: Converting raw item metadata into structured, machine-readable formats.
- **Feature Encoding**: Representing item features as dense vector embeddings, facilitating efficient retrieval, clustering, and downstream tasks.

2.1.1.1 Content Understanding Tasks

Key methodologies used in content understanding include:

- **Topic Classification**: Categorizing items according to predefined taxonomies, which structure content into hierarchical themes. The IAB (Interactive Advertising Bureau) Content Taxonomy is a standardized classification system developed by the IAB Tech Lab to categorize digital content. It provides a hierarchical structure that defines topics and subtopics across a wide range of

2.1 Recommendation System Workflow

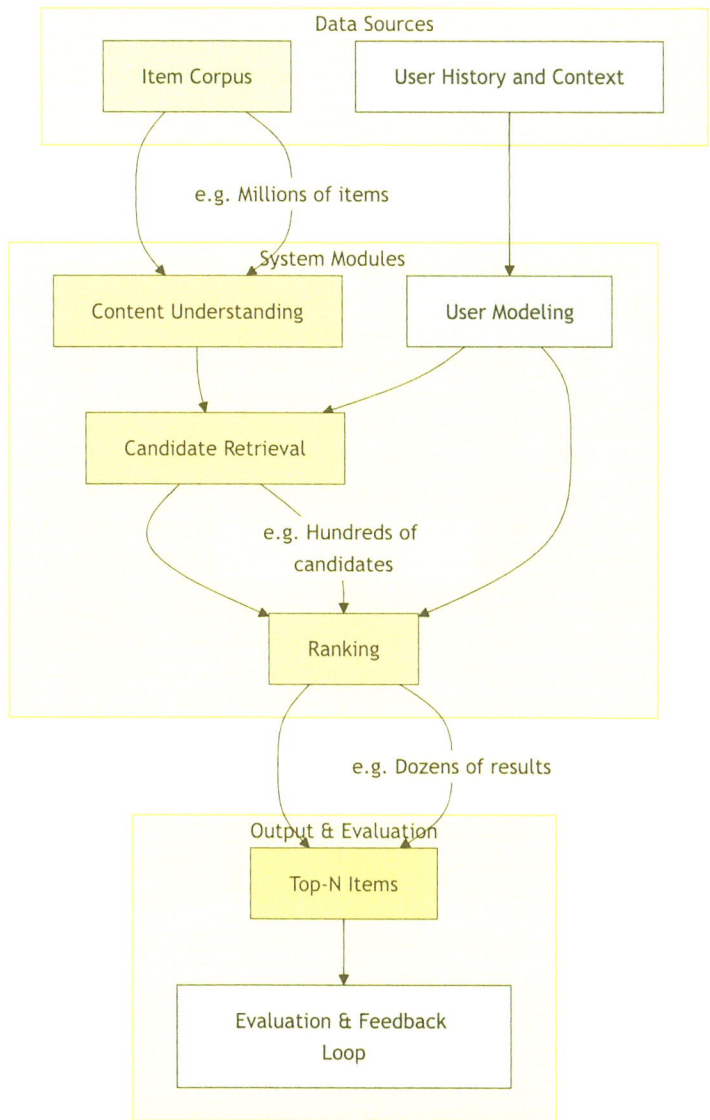

Fig. 2.1 Recommendation system workflow

domains (e.g., Arts, News, Technology, Sports). For instance, an article about casinos is classified under the hierarchy "Attractions → Casinos and Gambling." Taxonomies like these are widely used in both content-based filtering and supply-demand analysis within recommendation systems.
- **Entity Extraction**: Identifying and disambiguating key entities such as people, organizations, locations, and specific products within item descriptions. For

example, in a news article, extracting entities like "Apple" (the company) versus "apple" (the fruit) requires effective entity disambiguation. This step enhances contextual understanding and ensures recommendations are based on accurate associations, which is particularly useful in personalized search and discovery.
- **Sentiment Analysis**: Analyzing the sentiment expressed in user-generated content, such as reviews or comments, to determine the emotional tone associated with items. For example, sentiment analysis can extract positive or negative sentiments from user reviews of restaurants, tourist spots, or products, providing valuable insights into user perceptions that inform recommendation logic.
- **Key Phrase Extraction**: Identifying and extracting significant phrases that encapsulate the essential features or themes of an item. This process helps in summarizing content and improving search relevance by pinpointing the most relevant terms that describe an item's core attributes.
- **Content Quality**: Evaluating content for quality factors such as grammatical correctness, factual accuracy, and overall relevance. For example, content quality assessment can flag problematic characteristics like clickbait, misleading headlines, or disallowed traits such as violence or explicit content, ensuring that recommendations align with platform guidelines and user expectations.

These techniques are essential for enhancing recommendation relevance, enabling content supply–demand analysis, and supporting a healthier content ecosystem, and they are explored in greater depth in later chapters.

2.1.1.2 Content Understanding Methods

Content understanding in recommendation systems involves extracting meaningful information from various sources such as product descriptions, reviews, and user queries. Traditional methods for this task include the following:

1. **Classical Text Representation Methods (BoW, TF-IDF)**:
   - *Bag-of-Words (BoW)*: Represents text as sparse vectors of word occurrences, ignoring word order and relationships, which limits its ability to capture context.
   - *TF-IDF*: Adjusts word frequencies based on their importance within a corpus, emphasizing rarer terms, but fails to capture deeper semantic meaning or context.
2. **Pre-trained Word Embeddings**:
   - *Word2Vec*: Generates dense vector representations by modeling word co-occurrence in local contexts, capturing semantic relationships between words (Mikolov et al., 2013).
   - *GloVe*: Builds on Word2Vec by leveraging global co-occurrence statistics across the entire corpus to learn word embeddings, offering a more comprehensive view of word relationships (Pennington et al., 2014).

3. **Customized Models for Specific Tasks**: Traditional models were often task-specific, requiring separate models for tasks such as topic classification, key phrase extraction, and sentiment analysis. Examples of these models include:

   - *Topic Classification*: Models like Naive Bayes or Support Vector Machines (SVM) were often used for categorizing text into predefined topics.
   - *Key Phrase Extraction*: Unsupervised techniques such as TF-IDF and TextRank (Mihalcea & Tarau, 2004) have been widely adopted. TextRank applies graph-based ranking to identify salient terms within a document.
   - *Content Quality Modeling*: Classification models using rule-based systems or shallow learning algorithms were used to assess the quality of content, such as detecting clickbait or offensive content.

These task-specific models can now be replaced by LLM-based pipelines, which leverage large pre-trained language models to handle a range of content understanding tasks in a unified manner.

2.1.2 User Modeling

User modeling aims at understanding individual users and their preferences, by gathering data on user behavior, including past purchases, ratings, browsing history, and demographics. This data is used to create a user profile that reflects their interests and tendencies. Effective user modeling in recommender systems requires consideration of several key components:

- **Data Collection**: This involves gathering user interaction data (e.g., purchases, clicks, ratings) and contextual information (e.g., session time, location, device type). This data is captured through event tracking mechanisms while ensuring compliance with privacy regulations (e.g., GDPR, CCPA).
- **User Profile Construction**: Collected data is aggregated to form detailed user profiles, incorporating demographic details (e.g., age, location), behavioral patterns (e.g., browsing history), and contextual signals (e.g., device preferences). Profiles can be updated dynamically as user interests evolve over time. Key dimensions of user profiles:
  - *Demographic Information*: Age, gender, location, etc.
  - *User Interests*: Topics, genres, brands, etc.
  - *Behavioral History*: Browsing history, purchase history, click-through rates, etc.
- **User Modeling and Representation Learning**: User profiles are transformed into vector representations using techniques such as matrix factorization, deep neural models, or transformers, which capture latent preferences from sequential and contextual behaviors.

- *Embeddings* encode user preferences into dense vectors that allow for efficient similarity matching and downstream learning tasks.
- *Social Graph Models* leverage user–user relationships to model influence and peer dynamics, capturing social signals that enhance personalization.
- *Domain Ontologies* introduce structured external knowledge (e.g., product hierarchies, genre relationships) to improve generalization and semantic understanding in user models.

User modeling techniques fall into three categories: explicit feature-based models, implicit behavior models, and sequence models. We explain the three categories of models here and will present the most popular user modeling techniques like collaborative filtering, matrix factorization later.

- *Explicit Feature-Based Models* rely on structured, manually curated user attributes—such as age, gender, and location—as well as explicit user feedback like star ratings or labeled preferences. These models assume that user intent can be inferred from clearly defined features. Classical techniques include Logistic Regression and Decision Trees, which use demographic and transactional data to predict preferences. Matrix Factorization models, when applied to explicit ratings, decompose user-item matrices into latent factors representing user and item traits. While interpretable and efficient, these models struggle to capture subtle tastes (e.g., a preference for "dark comedies" over generic "comedies") and are limited in adapting to evolving or contextual user behaviors.
- *Implicit Behavior Models*, by contrast, infer preferences from user actions—clicks, views, dwell time, or purchases—without requiring explicit ratings or feedback. These models operate under the assumption that user behavior reflects latent intent, even if preferences are not directly stated. Collaborative filtering is a central technique, identifying patterns in co-interactions across users or items. *Factorization Machines* (Rendle, 2010) enhance these methods by modeling higher order feature interactions in sparse datasets. Hybrid models further combine behavioral signals with content metadata (e.g., item descriptions). While strong in capturing behavioral trends, these models often overlook unstructured or context-rich inputs—such as free-text queries or natural language reviews—that can provide deeper insights into user intent.
- *Sequence Models* extend implicit modeling by capturing the *temporal dynamics* of user behavior. Recurrent Neural Networks (RNNs) and Gated Recurrent Units (GRUs), such as GRU4Rec, specialize in session-based recommendation by predicting the next likely interaction. Transformer-based architectures like BERT4Rec (Sun et al., 2019) and SASRec (Kang & McAuley, 2018) improve upon these by modeling long-term dependencies using self-attention mechanisms. With the advent of LLMs, auto-regressive models like GPT-3/4 have been employed to generate recommendations directly from user dialogue or textual histories. While these models offer strong performance in capturing evolving preferences, they require large volumes of sequential data and significant compute resources, making them less suitable for sparse or cold-start scenarios.

2.1 Recommendation System Workflow

2.1.2.1 Collaborative Filtering (CF)

Collaborative Filtering (CF) is a foundational technique that recommends items based on user-item interaction patterns, without requiring explicit content features (Sarwar et al., 2001). There are two main variants:

- *User-Based CF* computes similarity (e.g., cosine or Pearson correlation) between user vectors and recommends items favored by similar users (Resnick et al., 1994).
- *Item-Based CF* computes similarity between item vectors based on co-occurrence or interaction patterns, recommending items similar to those the user has already liked.

Strengths

- Simple, interpretable, and effective for dense interaction matrices.
- Requires no domain-specific knowledge.

Limitations

- Suffers from data sparsity and cold-start issues.
- Similarity matrix computation scales poorly with large user/item sets.

2.1.2.2 Matrix Factorization (MF)

Matrix Factorization (MF) models user-item interactions by learning low-dimensional latent representations. It approximates the interaction matrix $R \in \mathfrak{R}^{m \times n}$ as the product of a user matrix $U \in \mathfrak{R}^{m \times k}$ and item matrix $V \in \mathfrak{R}^{n \times k}$, such that $R_{ij} \approx U_i^T V_j$ (Koren et al., 2009, Fig. 2.2).

Popular Algorithms

- **Singular Value Decomposition**: Factorizes RR using singular value decomposition on observed ratings.

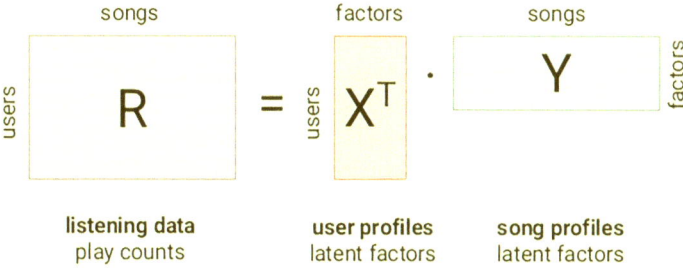

Fig. 2.2 Matrix factorization applied user listening data

- **Alternating Least Squares**: Alternates between fixing user and item factors to minimize squared error using least squares.

Strengths

- Handles sparse data and reveals latent dimensions (e.g., genre or style).
- Scalable with optimizations like parallelized ALS.

Limitations

- Assumes linear interaction, missing complex relationships.
- Requires retraining to incorporate new users/items.

2.1.2.3 Factorization Machines (FM)

Factorization Machines (FM) generalize matrix factorization by modeling *pairwise interactions between arbitrary features* (e.g., user age, item genre, context). For an input feature vector $x \in \Re^n$, FM models pairwise interactions as:

$$\hat{y}(x) = w_0 + \sum_{i=1}^{n} w_i x_i + \sum_{i=1}^{n}\sum_{j=1}^{n} v_i, v_j x_i x_j$$

where $v_i \in \Re^k$ are latent vectors capturing interaction effects (Rendle, 2010).

Strengths

- Captures high-order interactions across sparse, high-dimensional inputs.
- Versatile for tasks like CTR prediction and hybrid recommendation.

Limitations

- Manual feature engineering is often required.
- Training can be slow for large input spaces due to interaction expansion.

2.1.3 Candidate Retrieval

Imagine designing a recommender system for a platform with millions of items—books, movies, products, or articles. Scoring every item in real time is computationally infeasible. This is where *candidate retrieval* plays a critical role: it filters the catalog down to a small, high-recall pool of relevant items tailored to the user, ensuring downstream ranking models only evaluate promising candidates. To achieve this, multiple retrieval paths are often employed, such as:

- **Item-Based Retrieval**: Retrieving items similar to those the user has previously liked or interacted with (e.g., recommending "Inception" to a user who enjoyed "The Dark Knight").

2.1 Recommendation System Workflow 49

- **Topic-Based Retrieval**: Retrieving items related to topics the user has shown interest in (e.g., suggesting articles on "machine learning" for a user who follows AI-related content).
- **User-Based Collaborative Filtering**: Leveraging data from similar users to recommend items based on collective preferences.
- **Hybrid Approaches**: Combining multiple techniques to balance accuracy, diversity, and coverage.

At its core, candidate retrieval focuses on *recall over precision*, aiming to capture all potentially relevant items. In traditional systems, this stage relies on heuristic, statistical, or collaborative methods. Modern approaches, however, increasingly leverage learned representations and semantic understanding powered by deep learning and more recently, LLMs.

2.1.3.1 Content-Based Retrieval

Content-based methods recommend items by analyzing the *attributes or features of the items themselves*, matching them with user preferences derived from past interactions.

- **Keyword Matching**: Matches user queries or item preferences based on exact term overlap.
    - *Strength*: Fast and easy to implement.
    - *Limitation*: Limited expressiveness and poor semantic understanding.
- **TF-IDF (Term Frequency-Inverse Document Frequency)**: Evaluates term importance in item descriptions or user queries.
    - *Strength*: Effective for sparse or domain-specific textual data.
    - *Limitation*: Fails to capture synonyms or semantics.
    - *Example*: A search for "wireless headphones" returns items with exact terms but misses "Bluetooth earbuds."
- **Cosine Similarity**: Measures the similarity between feature vectors (e.g., TF-IDF or embedding-based).
    - *Strength*: Efficient for content matching.
    - *Limitation*: Heavily reliant on representation quality.

2.1.3.2 Collaborative Filtering Retrieval

Collaborative filtering generates recommendations using *user-item interaction patterns*, without requiring explicit item attributes.

- **User-Based Collaborative Filtering**: Recommends items preferred by users with similar behavior.
  - *Strength*: Simple and interpretable.
  - *Limitation*: Struggles with sparse or new user data.
- **Item-Based Collaborative Filtering**: Recommends items that tend to co-occur in user interactions.
  - *Example*: If many users who liked "Item X" also liked "Item Y," the system recommends "Item Y."
  - *Strength*: More scalable for stable catalogs.
  - *Limitation*: Requires significant interaction data.
- **Matrix Factorization (e.g., ALS)**: Projects users and items into a shared latent space using past interaction data.
  - *Strength*: Learns abstract user/item preferences for efficient retrieval.
  - *Limitation*: Cold-start and interpretability remain challenges.

2.1.3.3 Neural Retrieval

Neural methods generate dense embeddings for text or other modalities, improving semantic matching (Covington et al., 2016). Early dense retrieval models like Microsoft's Deep Structured Semantic Model (DSSM) (Huang et al., 2013) introduced the use of deep neural networks to map users and items into a shared embedding space for semantic matching. While effective at improving retrieval relevance, DSSM used static embeddings and lacked adaptability to user context.

Two-Tower Neural Network (TTSN)

The Two-Tower Neural Network (TTSN) architecture improves scalability and flexibility by separating the user and item modeling pipelines. As shown in the left panel of Fig. 2.3, user features (e.g., profile, past interactions) are processed through a user tower, while item features go through a separate item tower. Both outputs are projected into a shared embedding space. The final recommendation score is typically computed using a *dot product* or *similarity function* between the user and item embeddings (Yi et al., 2019).

- **Strengths**: TTSNs allow independent pre-training and indexing of item embeddings, making them efficient for large-scale retrieval tasks. They also capture non-linear relationships between user/item attributes.
- **Limitations**: They struggle to capture fine-grained user intent in context-rich or sequential settings, especially when user preferences evolve quickly.

2.1 Recommendation System Workflow

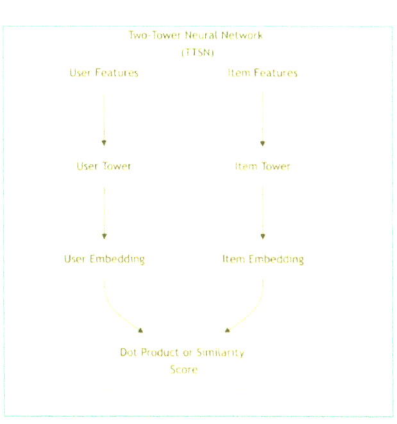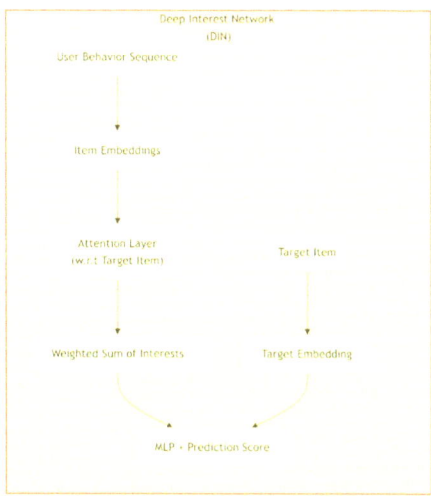

Fig. 2.3 Model architectures for TTSN and DIN

Deep Interest Network (DIN)

To address the limitation of fixed embeddings, the Deep Interest Network (DIN) introduces attention-based mechanisms that dynamically adjust user representations based on the target item (Zhou et al., 2018). As illustrated in the right panel of Fig. 2.3, DIN uses a user's historical behavior sequence and applies an attention layer to weigh past item embeddings according to their relevance to the current target item. The weighted sum is then combined with the target item's embedding and passed through a multilayer perceptron (MLP) to compute the final prediction score.

- **Strengths**: DIN enables *context-aware personalization*, adapting to the user's intent in real time by focusing on the most relevant behavioral signals.
- **Limitations**: It introduces higher *computational overhead* during training and inference, especially for long user histories.

To scale these dense retrieval systems in practice, Chap. 3 will introduce Approximate Nearest Neighbor (ANN) algorithms such as LSH, ANNOY, and HNSW, which enable efficient retrieval from large embedding spaces.

2.1.4 Ranking

Ranking is the process of prioritizing candidate items based on their predicted relevance to the user. This stage typically employs *heavier models* to predict user engagement and optimize for specific objectives. Key components include:

- **Relevance Prediction**:
  - A multi-task model predicts various engagement signals, such as the likelihood of a user clicking, liking, saving, or commenting on an item.
  - These predictions are combined into a composite score that reflects the item's overall relevance to the user.
- **List-Level Optimization**:
  - After initial scoring, a *reranking step* may be applied to optimize the list as a whole, incorporating objectives such as diversity, fairness, or business goals (e.g., promoting new or high-margin items).
  - Techniques like determinantal point processes (DPPs) or learned Rerankers are often used to balance relevance with diversity.

Effective ranking ensures that users are presented with the most pertinent and engaging items in a clear and actionable manner, ultimately driving higher user satisfaction and platform engagement.

2.1.4.1 Traditional Ranking Methods

After retrieving candidate items, ranking determines their order based on relevance to the user, significantly influencing engagement. Traditional methods often used linear models or pointwise approaches, relying on features like user-item interactions or item popularity to compute relevance scores. While simple and interpretable, these methods struggled to capture complex user preferences.

Machine learning introduced pairwise and listwise approaches, leveraging algorithms like gradient boosting and neural networks to model feature interactions more effectively. In this section, we will start from pointwise ranking and then move on to discuss pairwise and listwise ranking.

Pointwise Ranking

Pointwise ranking methods predict the relevance of each individual item in relation to the user's query or history. The system ranks the items one by one based on predicted scores.

- **Logistic Regression or SVMs**:
  - These classifiers were used to predict whether a given item is relevant or not, given a user-query pair (pointwise). For example, whether a specific product is relevant to a user can be predicted using these models based on user features, item features, and their interactions.
  - The model would assign a score to each item, and items with higher scores would be ranked higher.

2.1 Recommendation System Workflow

Pairwise Ranking

Pairwise ranking methods compare two items at a time to determine which one is more relevant to the user (Burges et al., 2005). These methods are based on ranking pairs of items rather than individual items.

- **RankNet**:
  - A neural network-based approach that compares pairs of items and learns to rank one over the other (Burges et al., 2005). The model predicts which item is more relevant given a pair and adjusts weights based on the ranking error.
  - This method focuses on the *relative* order between items, rather than their absolute relevance scores.
- **SVM-Rank**:
  - A support vector machine (SVM) approach used to learn ranking models based on pairwise comparisons (Joachims, 2006). Similar to RankNet, SVM-Rank optimizes the order of items by minimizing ranking errors across pairs.

Listwise Ranking

Listwise ranking methods evaluate an entire list of items simultaneously (Cao et al., 2007). These methods rank multiple items at once, considering their positions relative to each other within a list.

- **ListNet**: ListNet uses a probabilistic model to rank a list of items (Cao et al., 2007). The goal is to optimize the probability distribution of the list order rather than focusing on individual pairwise comparisons.
- **LambdaRank**: An extension of RankNet, LambdaRank optimizes the ranking function by focusing on the *gradients* of ranked lists, using *lambda* values to fine-tune the ranking performance (Burges, 2010). It's particularly effective in optimizing large-scale ranking tasks.
- **LambdaMART**: Combines LambdaRank's listwise gradient optimization with gradient-boosted decision trees (MART) for more powerful non-linear modeling. LambdaMART is widely used and competitive learning-to-rank algorithms in production settings due to its robustness, interpretability, and strong empirical performance (Burges, 2010).

2.1.4.2 Traditional Reranking Methods

After an initial list of recommended items is retrieved, *reranking* refines the results to better serve personalization, diversity, or business objectives. Unlike the initial ranking which optimizes general relevance, reranking integrates additional criteria through rules, heuristics, or post-processing steps.

Manual and Heuristic-Based Reranking

This category includes both *rule-based* and *heuristic* approaches that adjust rankings using predefined or adaptive criteria:

- **Boosting and Demotion**: Promote items with desirable attributes like popularity, recency, or newness; demote stale or over-recommended items.
- **Diversity Control**: Rearrange similar items to avoid redundancy and improve list heterogeneity.
- **Personalization Adjustments**: Elevate items aligned with a user's historical preferences or inferred interests.
- **Contextual Heuristics**: Use local signals such as device type, time of day, or recent activity to refine the list dynamically.

Post-processing and Business Constraints

These techniques refine the output ranking to comply with *external constraints or commercial priorities*:

- **Sponsored Item Placement**: Ensure priority positioning for promoted or paid content.
- **Category Quotas**: Enforce diversity by limiting over-representation of specific item types.
- **Contextual Reweighting**: Apply localized tweaks (e.g., geolocation-based adjustments) after model scoring.

Large Language Models (LLMs) enhance recommendation system ranking and reranking in several ways:

- **Embeddings as Ranking Features**. LLM-generated embeddings from item descriptions, queries, or user profiles can be used as input features for ranking models. They capture semantic relationships beyond traditional collaborative signals.
- **Prompt-Based Ranking**. LLMs can be prompted to compare and rank items directly, enabling zero-shot or few-shot ranking without model retraining.
- **Synthetic Training Data Generation**. LLMs can simulate user preferences by generating pairwise comparisons or relevance labels, improving data efficiency for supervised ranking.
- **LLM-as-a-Judge for Evaluation**. LLMs can assess ranked lists by judging relevance, fluency, or personalization quality, offering scalable evaluation without full user studies.

2.1.5 Evaluation

Evaluation is an essential component of recommendation systems, providing a framework for assessing their performance and optimizing for better user outcomes (Jannach et al., 2011). As illustrated in Fig. 2.4, recommendation evaluation metrics can be broadly classified into three categories:

2.1.5.1 Business Metrics

Business metrics quantify the real-world impact of a recommendation system on organizational goals, such as user engagement, retention, and revenue growth (Gunawardana & Shani, 2015). These metrics are crucial for aligning machine learning performance with business outcomes.

- **Click-Through Rate (CTR)**: Measures the ratio of clicks to impressions, indicating how often users interact with recommended items. It is formally defined as:

$$CTR = \frac{\text{Number of clicks}}{\text{Number of impressions}}$$

In advertising-driven models, optimizing CTR is critical for maximizing ad revenue (McMahan et al., 2013). A/B testing frameworks are often employed to compare recommendation strategies by measuring statistically significant differences in CTR.

- **Conversion Rate (CVR)**: Tracks the percentage of users who complete a desired action (e.g., purchase, subscription) after interacting with a recommendation:

$$CVR = \frac{\text{Number of conversions}}{\text{Number of clicks}}$$

Multi-touch attribution models help determine how recommendations contribute to conversions across user sessions.

Fig. 2.4 Evaluation metrics for recommendation systems

- **Gross Merchandise Value (GMV)**: Represents the total sales volume generated through recommendations, often used in e-commerce to assess revenue impact.

$$\text{GMV} = \sum_{i=1}^{N} \text{Price}_i \times \text{Quantity}_i$$

  where N is the number of transactions influenced by recommendations.
- **Customer Lifetime Value (CLTV)**: Estimates the long-term revenue contribution of a user, factoring in retention improvements from personalized recommendations (Gupta et al., 2006).

These metrics are particularly relevant in *advertising-driven models* (where engagement directly impacts ad revenue) and *subscription-based models* (where retention and churn reduction are key).

2.1.5.2 Model Metrics

Model metrics evaluate the predictive and ranking performance of recommendation algorithms, ensuring they accurately match user preferences.

Rating Prediction Metrics

For explicit feedback (e.g., star ratings), regression-based metrics are used:

- **Root Mean Squared Error (RMSE)**:

$$\text{RMSE} = \sqrt{\frac{1}{N} \sum_{i=1}^{N} (y_i - \hat{y}_i)^2}$$

  Penalizes large errors more severely due to the squared term.
- **Mean Absolute Error (MAE)**:

$$\text{MAE} = \frac{1}{N} \sum_{i=1}^{N} |y_i - \hat{y}_i|$$

  More interpretable but less sensitive to outliers.

Classification and Ranking Metrics

For implicit feedback (e.g., clicks, purchases), ranking quality is critical:

- **Precision@K**: Fraction of relevant items in the top-K recommendations.

2.1 Recommendation System Workflow

$$\text{Precision@K} = \frac{|\text{Relevant items} \cap \text{Top-K recommendations}|}{K}$$

- **Recall@K**: Fraction of all relevant items captured in the top-K.

$$\text{Recall@K} = \frac{|\text{Relevant items} \cap \text{Top-K recommendations}|}{|\text{All relevant items}|}$$

- **Mean Average Precision (MAP@K)**: Extends Precision@K by averaging over all user queries, rewarding systems that rank relevant items higher.

$$AP@K = \frac{1}{m}\sum_{k=1}^{K} \text{Precision}@k \times \text{rel}(k)$$

where (rel(k)) is an indicator function for relevance at rank (k), and m is the number of relevant items.

- **Normalized Discounted Cumulative Gain (NDCG@K)**: Measures ranking quality with graded relevance (e.g., strongly vs. weakly preferred items).

$$NDCG@K = \frac{DCG@K}{IDCG@K}, \quad DCG@K = \sum_{i=1}^{K} \frac{2^{\text{rel}_i} - 1}{\log_2(i+1)}$$

where (IDCG@K) is the ideal DCG for perfect ranking.

These metrics are essential for optimizing *collaborative filtering*, *matrix factorization* and *neural recommendation models* (Koren et al., 2009; Rendle et al., 2020).

2.1.5.3 Outcome Metrics

Beyond accuracy, recommendation systems must ensure diversity, novelty, and fairness to enhance user satisfaction (Shani & Gunawardana, 2011).

- **Coverage**: Measures the fraction of items the system can recommend.

$$\text{Coverage} = \frac{|\text{Recommended items}|}{|\text{Total items}|}$$

Low coverage indicates a "rich-get-richer" bias, where only popular items are recommended.

- **Diversity**: Quantifies dissimilarity between recommended items, often using intra-list distance (Ziegler et al., 2005):

$$\text{Diversity} = 1 - \frac{1}{K(K-1)} \sum_{i \neq j} \text{sim}(i,j)$$

where (sim(i, j)) is a similarity metric (e.g., cosine similarity in embedding space).
- **Novelty and Serendipity**:
  - **Novelty**: Measures how unfamiliar recommended items are to users.
  - **Serendipity**: Balances relevance and unexpectedness.
- **Fairness**: Ensures equitable exposure across item providers or demographic groups (Mehrotra et al., 2018). Common fairness metrics include *demographic parity* and *equal opportunity*.

These metrics help mitigate *filter bubbles* and improve long-term user engagement by balancing exploration-exploitation trade-offs.

Together, these metrics help optimize recommendation systems by balancing business goals, improving user satisfaction, and ensuring diverse, engaging recommendations.

The evaluation process typically follows a three-phase approach:

- **Offline Evaluation**: Using historical data and model metrics for initial testing and adjustments before deployment.
- **A/B Testing**: Conducting live, controlled experiments to measure the system's effectiveness in real-world settings.
- **Continuous Monitoring**: Continuously refining and improving the system based on ongoing feedback, ensuring sustained user satisfaction and system relevance.

This structured approach ensures that recommendation systems are rigorously tested and continuously optimized for both performance and user satisfaction.

2.2 Challenges and Transition to LLM-Powered Systems

Traditional recommendation systems have been widely adopted for their simplicity and effectiveness in leveraging user-item interaction data. However, they face significant limitations in handling the complexity, diversity, and dynamism of modern recommendation tasks. These challenges can be categorized into three levels: user-level, item-level, and model-level. Below, we outline these challenges and highlight how they motivate the transition to LLM-powered recommendation systems.

2.2 Challenges and Transition to LLM-Powered Systems

2.2.1 User-Level Challenges

Traditional systems struggle to model the diversity and scale of user behavior, particularly on platforms with hundreds of millions of users. User preferences are influenced by many external, often unobservable factors—social trends, emotional states, and context (e.g., time, location). For example:

- A user might shop for seasonal products (e.g., winter coats in December) or make decisions driven by real-time events (e.g., buying fitness gear after New Year's).
- Emotional states can drive preferences, such as binge-watching comedies when stressed.
- Complex natural language queries (e.g., "action-packed but family-friendly movies from the last decade") are difficult for traditional systems that lack semantic understanding.

Limitations of Traditional Methods
Collaborative filtering models rely on static user-item matrices, which fail to account for temporal shifts or incorporate external signals like trending topics. These systems also struggle to interpret unstructured data such as reviews or queries.

LLM Opportunity
Large Language Models (LLMs) address these issues by unifying behavioral, textual, and contextual data in a coherent semantic space. They model dynamic preferences using real-time cues and language-based reasoning, enabling more adaptive recommendations (Zhang et al., 2023; Wu et al., 2023).

2.2.2 Item-Level Challenges

At the item level, traditional methods often fail to capture niche or context-specific relationships—especially for items with low interaction frequency (long-tail items). For example:

- A user purchasing both a yoga mat and a fitness tracker shares a wellness intent though these items rarely co-occur.
- Users may consume diverse genres depending on mood or situation (e.g., switching between documentaries and comedies).

Limitations of Traditional Methods
Matrix factorization and other collaborative filtering techniques depend on co-occurrence, which doesn't capture semantic or multi-modal connections. Rich textual descriptions, reviews, and images are often ignored.

LLM Opportunity
LLMs can infer latent themes from unstructured content, bridging the semantic gap between items with low observed similarity. For instance, models like M6-Rec and IDGenRec use language understanding to align item representations through metadata or learned textual identifiers (Cui et al., 2022; Tan et al., 2024).

2.2.3 Model-Level Challenges

Traditional systems face difficulties generalizing to new users, items, or emerging content:

- **Cold-Start Problem**: New users or items lack interaction history.
- **Data Sparsity**: Sparse interactions hinder personalization.
- **Scalability**: Matrix-based models become inefficient at scale.

Limitations of Traditional Methods
These models cannot infer preferences without interaction data and lack mechanisms for incorporating rich content. They also require retraining to adapt to updates, which is computationally expensive.

LLM Opportunity
LLMs can perform zero-shot reasoning over new items by interpreting their content. Techniques such as prompt tuning, PEFT (e.g., LoRA, QLoRA), and RLHF enhance performance while maintaining scalability (Wu et al., 2023; Kim et al., 2024). Hybrid methods like A-LLMRec combine collaborative filtering with LLMs to improve generalization and reduce cold-start issues (Kim et al., 2024).

2.2.4 Other Challenges and LLM Opportunities

- **Semantic Gap**: Traditional models fail to capture the meaning embedded in text, reviews, or product descriptions.
- **Explainability**: ID-based models cannot explain why items are recommended, reducing user trust.

LLM Opportunity
LLMs offer semantic understanding, explainability, and flexibility:

- Parse complex queries and generate structured outputs (e.g., "Find budget-friendly romantic comedies").
- Unify multi-modal data (text, images, interactions) for deeper insights (Zhang et al., 2023).
- Support interaction and reasoning in recommendation via natural language (He et al., 2023).

Table 2.1 Challenges of traditional methods and the respective LLM solution

Challenge category	Traditional method limitations	LLM-powered solutions
User-level challenges	Dynamic preferences External factors	Contextual understanding Dynamic adaptation
Item-level challenges	Niche relationships Sparse data	Semantic bridging Cross-modal alignment
Model-level challenges	Cold-start problem Sparse interaction data Scalability	Zero-shot recommendations Synthetic data generation Efficient inference
Additional limitations	Lack of contextual understanding Semantic gap	Semantic parsing Unified data representation

Table 2.1 summarizes the traditional challenges and LLM solutions. These challenges underscore the need for recommendation paradigms that transcend interaction-based modeling. LLM-powered systems address these gaps by:

- *Unifying diverse data sources* (e.g., behavioral, textual, contextual) into a shared representation space.
- *Enabling dynamic adaptation* to evolving user preferences and real-time events.
- *Bridging the semantic gap* through cross-modal understanding and zero-shot generalization.
- *Improving scalability and efficiency* via lightweight fine-tuning and embedding-based retrieval.

2.3 LLMs Paradigms in Recommendation Systems

LLM-powered recommendation systems can be broadly categorized into two paradigms: *LLM-enhanced recommendation systems* and *LLM as recommendation systems*. These paradigms differ in their design philosophy, implementation complexity, and practical trade-offs. Below, we explore their key differences, use cases, and considerations for choosing between them. The choice between these paradigms depends on both *design considerations* and *practical constraints* (Table 2.2):

2.3.1 LLM-Enhanced Recommendation Systems

LLM-enhanced systems integrate large language models into existing recommendation architectures, primarily as powerful feature extractors or auxiliary modules. Instead of replacing the recommendation engine, LLMs enhance it through semantic embeddings, contextual signals, or token-level representations.

Table 2.2 Comparison between two paradigms of LLM-powered recommendation

Aspect	LLM as enhancer	LLM as recommender
Role of LLM	Augments traditional models by generating embeddings or tokens, labeling data for training and evaluation	Directly generates recommendations from user profiles and prompts
Complexity	Easier to integrate into existing pipelines	Requires end-to-end adaptation of LLMs for recommendation tasks
Costs	Lower computational overhead; leverages existing infrastructure	Higher computational costs due to LLM inference; requires prompt engineering
Ideal for	Enhancing specific components (e.g., item representation)	End-to-end personalization in dynamic or conversational settings

Key Advantages

- **Modular Integration**: LLMs can be plugged into legacy systems, allowing organizations to reuse existing infrastructure.
- **Richer Representations**: LLM-derived embeddings encode deep semantic and contextual signals from unstructured text (e.g., reviews, descriptions, queries).
- **Low-Friction Deployment**: Enhancements like feature enrichment or reranking can be introduced without full system overhauls.

Example Workflow

- **Input**: Item corpus and user history (e.g., past movie ratings and descriptions).
- **Processing**: Use LLM to generate item or user embeddings and augment features for candidate retrieval or ranking.
- **Output**: Refined ranking scores incorporating semantic similarity and context-aware features.

Challenges

- **Latency and Cost**: LLM inference can increase runtime and resource usage.
- **Alignment with Objectives**: Extracted features must align with the recommendation task (e.g., CTR, NDCG).
- **Scalability**: Embedding large item corpora with LLMs requires efficient batching and storage strategies.

2.3.2 LLM as Recommendation Systems

In this paradigm, pre-trained LLMs directly serve as the recommendation engine (He et al., 2023). User data is input as structured prompts, and the LLM generates outputs like itineraries or dining suggestions.

Key Advantages

- **End-to-End Personalization**: LLMs process context-rich data (e.g., real-time preferences).

- **Conversational Capabilities**: Enable interactive refinement (e.g., "Find cheaper options").

Example Workflow

- **Input**: "Family-friendly beach resorts in Europe under $200/night."
- **Processing**: LLM generates hotel, activity, and dining recommendations.
- **Output**: A tailored travel plan with budget-aware options.

Challenges

- **Computational Costs**: Requires optimization via distillation.
- **Prompt Engineering**: Critical for relevance.

2.3.3 Practical Considerations

The choice between using LLM as Enhancer or LLM as Recommender depends on several factors, including the system's complexity, resource constraints, and the specific goals of the recommendation process.

LLM as Recommender is simpler in design, eliminating the need for multiple models, but comes with higher computational costs, fine-tuning requirements, and potential opacity. It is ideal for autonomous, context-aware recommendations, like personalized travel planning, but requires significant resources and careful prompt design.

LLM as Enhancer integrates with existing recommendation systems, enhancing them with richer features and embeddings. It's more cost-effective and preserves flexibility, making it suitable for scenarios where you want to improve recommendation quality without overhauling the architecture.

Key Considerations

- **LLM as Enhancer**: Best for integrating LLMs into existing systems with minimal disruption and lower cost.
- **LLM as Recommender**: Ideal for high personalization or conversational recommendations, but requires more resources and fine-tuning.

Practical Tips

- **Infrastructure**: If you have an existing system, LLM as Enhancer may be more efficient and cost-effective.
- **Resources**: LLM as Recommender needs substantial computational power, so consider LLM as Enhancer if resources are limited.
- **Goals**: For personalized, conversational recommendations, choose LLM as Recommender; for general recommendations, LLM as Enhancer may be sufficient.

In Chaps. 3 and 4, we will dive deeper into LLM-enhanced and LLM-based recommendation systems.

2.4 Tutorial: From Traditional to LLM-Based Recommendations Using MovieLens Dataset

2.4.1 Overview

This tutorial presents a lightweight yet illustrative study comparing traditional collaborative filtering (CF) and large language model (LLM)-based prompting methods for movie recommendation. We demonstrate how to design experiments, construct prompts, and assess recommendation quality through standard evaluation metrics.

Goal of This Tutorial

- Understand the experimental design for evaluating recommendation systems.
- Learn how to structure LLM prompts for recommendation tasks.
- Evaluate and compare recommendation systems using multifaceted metrics (accuracy, diversity, coverage) and practical considerations.

We show a condensed version of this tutorial in the book text. The full version of the code is available at: https://github.com/qqwjq1981/springer-LLM-recommendation-system

2.4.2 Experimental Design

2.4.2.1 Dataset and Train-Test Split

The experiments utilize the *MovieLens ml-1m dataset*, which contains structured records of user-item interactions, including movie titles and genre metadata. To emulate real-world recommendation scenarios, we employ a temporal split strategy, reserving the earliest 90% of each user's interactions for training and the most recent 10% for testing. This approach ensures that the model is evaluated on its capacity to generalize to future user preferences based on historical interaction patterns.

2.4.2.2 Methods Compared

The study compares two representative methods:

- **Collaborative Filtering (CF)**: A matrix factorization model (SVD) that learns latent representations from user-item ratings and predicts unseen preferences.
- **LLM-Based Prompting with Movie Titles**: An approach that constructs natural language prompts incorporating tokenized user IDs, liked and disliked movie

2.4 Tutorial: From Traditional to LLM-Based Recommendations Using MovieLens... 65

titles, and queries the LLM to generate recommendations. Only movie titles (not genres or metadata) are included for simplicity and interpretability.

2.4.2.3 Prompt Design

The LLM-based method uses structured prompts of the form:

```
You are a helpful movie recommendation assistant.
The user USER_1680 liked the following movies: One Flew Over
the Cuckoo's Nest (1975),
Miracle on 34th Street (1947), Airplane! (1980), Bambi
(1942), Sixth Sense, The (1999),
Run Lola Run (Lola rennt) (1998), Dumbo (1941), Saving
Private Ryan (1998), Fargo (1996), Verdict, The (1982).
The user USER_1680 disliked these movies: Pocahontas (1995),
Outbreak (1995), Bodyguard, The (1992),
Braveheart (1995), Like Water for Chocolate (Como agua para
chocolate) (1992), Man in the Iron Mask, The (1998),
Armageddon (1998), Conspiracy Theory (1997), Chariots of
Fire (1981), Young Guns (1988).
Please recommend exactly 5 movies that are similar to the
liked ones and different from the disliked ones.
Output only the recommended movie titles separated
by commas.
# output A Few Good Men, The Shawshank Redemption, The Green
Mile, Good Will Hunting, The Princess Bride
```

This format encourages the model to leverage semantic associations in the movie titles to generate plausible recommendations.

2.4.2.4 Inference Model

To ensure accessibility and responsiveness, the experiment uses *GPT-4o-mini API*, which provides a balance between semantic capabilities and inference latency. Earlier trials with GPT-2 showed limitations in quality and relevance of generation, reinforcing the need for more capable models.

2.4.2.5 Evaluation Metrics

The performance of the Traditional Collaborative Filtering (CF) model and the LLM Prompt-Based Model (using movie titles only) was evaluated using the following metrics:

1. **Recall@k**: Measures the proportion of relevant items in the top-k recommendations.
2. **Precision@k**: Measures the proportion of top-k recommendations that are relevant.
3. **NDCG@k**: Evaluates the ranking quality of the top-k recommendations.
4. **Catalog Coverage Ratio**: Measures the fraction of the item catalog that is recommended.
5. **Entropy Diversity**: Quantifies the diversity of recommendations using entropy.
6. **Execution Time**: Time taken to generate recommendations (in seconds).

2.4.3 Results and Analysis

2.4.3.1 Results Summary

Metric	k	CF	LLM	Metric	k	CF	LLM
Recall@k	5	0.019	0.009	Precision@k	5	0.129	0.056
	10	0.035	0.015		10	0.121	0.050
	20	0.061	0.022		20	0.103	0.040
NDCG@k	5	0.271	0.139	Catalog coverage ratio	5	0.126	0.101
	10	0.309	0.165		10	0.175	0.135
	20	0.331	0.195		20	0.231	0.195
Entropy diversity	5	0.602	0.498	Execution time (s)	5	26.52	1175.09
	10	0.639	0.514		10	25.72	1651.42
	20	0.673	0.527		20	25.80	2825.59

Results indicate that:

1. **LLM Prompt-Based Model Shows Decent Zero-Shot Performance**: Despite using a general-purpose LLM without domain-specific fine-tuning, the LLM-based method achieves non-trivial recommendation quality across all metrics. This demonstrates the potential of out-of-the-box LLMs in recommendation tasks using only movie titles.
2. **Traditional CF Still Outperforms LLM in Core Metrics**: Across Recall@K, Precision@K, and NDCG@K, traditional CF performs significantly better. This highlights the effectiveness of interaction-based learning and the need for fine-tuning to bridge this gap in LLMs.

3. **Catalog Coverage and Diversity Trade-Off**: LLM-based recommendations tend to offer moderately diverse and broad coverage, albeit slightly lower than CF. However, the Catalog Coverage Ratio and Entropy Diversity are still respectable, suggesting LLMs don't overly concentrate recommendations on popular items.
4. **Scalability Remains a Challenge for LLMs:** The execution time of LLM prompting is orders of magnitude higher (e.g., ~28× slower at top-20). This highlights the latency bottleneck of generative methods and the need for distillation or lightweight retrieval-based alternatives for practical deployment.

2.4.3.2 Advanced Methods

The limitations observed in using general-domain LLM APIs for recommendation tasks highlight the need for more advanced approaches. These methods aim to address the challenges of context length, latency, and correctness while leveraging the strengths of LLMs:

1. **Handling Long User History**: Newer LLMs with extended context capabilities can accommodate longer user histories, enabling richer personalization. Additionally, *prompt compression* techniques, such as summarizing user history into higher level preference descriptors, and *constrained generation*, such as forcing output to adhere to predefined formats or item catalogs, help reduce output token consumption and increase recommendation efficiency. These techniques allow better utilization of both input and output tokens, enabling more scalable and precise recommendation using LLMs.
2. **Foundational Models for Recommendations**: Foundational models pre-trained on large-scale recommendation datasets (e.g., user interactions, item metadata) can be fine-tuned for specific tasks, providing a balance between generalization and domain specificity. These models can handle longer user histories and generate recommendations grounded in the actual item catalog.
3. **Fine-Tuning LLMs with Domain-Specific Data**: Fine-tuning LLMs on domain-specific preference data (e.g., movie ratings, reviews) allows the model to better understand the recommendation context and generate more accurate suggestions. This reduces the reliance on prompt engineering and mitigates issues like fabricated recommendations.
4. **Model Distillation**: Distilling large LLMs into smaller, more efficient models reduces inference costs and latency, making LLM-based recommendations feasible for real-time applications.
5. **Hybrid Modeling**: Combining traditional methods (e.g., CF) with LLMs leverages the strengths of both: the efficiency and robustness of CF and the expressiveness of LLMs. For instance, CF can handle user-item interactions, while LLMs can incorporate contextual information like reviews or genres.

2.4.4 Conclusions

This tutorial sets the foundation for deeper exploration into advanced methods that address the limitations of zero-shot LLM prompting. By fine-tuning LLMs, distilling models, and leveraging hybrid approaches, we can overcome challenges like context length limits, latency, and correctness while unlocking the full potential of LLMs for recommendation tasks. These advanced directions will be discussed further in later sections and serve as a natural extension beyond zero-shot prompting.

References

Burges, C. J. C. (2010). From RankNet to LambdaRank to LambdaMART: An overview. Microsoft Research Technical Report MSR-TR-2010-82.

Burges, C., Shaked, T., Renshaw, E., et al. (2005). Learning to rank using gradient descent. In *Proceedings of the 22nd International Conference on Machine Learning* (pp. 89–96).

Cao, Z., Qin, T., Liu, T. Y., et al. (2007). Learning to rank: From pairwise approach to listwise approach. In *Proceedings of the 24th International Conference on Machine Learning* (pp. 129–136).

Covington, P., Adams, J., & Sargin, E. (2016). Deep neural networks for youtube recommendations. In *Proceedings of the 10th ACM Conference on Recommender Systems* (pp. 191–198).

Cui, Z., Ma, J., Zhou, C., Zhou, J., & Yang, H. (2022). M6-rec: Generative pretrained language models are open-ended recommender systems. CoRR abs/2205.08084.

Gunawardana, A., & Shani, G. (2015). Evaluating recommender systems. In *Recommender systems handbook* (pp. 265–308). Springer.

Gupta, S., Hanssens, D., Hardie, B., Kahn, W., Kumar, V., Lin, N., et al. (2006). Modeling customer lifetime value. *Journal of Service Research, 9*(2), 139–155. https://doi.org/10.1177/1094670506293810

He, Z., Xie, Z., Jha, R., Steck, H., Liang, D., Feng, Y., Majumder, B. P., Kallus, N., & McAuley, J. (2023). *Large language models as zero-shot conversational recommenders*. arXiv preprint arXiv:2308.10053.

Huang, P. S., He, X., Gao, J., et al. (2013). Learning deep structured semantic models for web search using clickthrough data. In *Proceedings of the 22nd ACM International Conference on Information & Knowledge Management* (pp. 2333–2338).

Jannach, D., Zanker, M., Felfernig, A., et al. (2011). *Recommender systems: An introduction*. Cambridge University Press.

Joachims, T. (2006). Training linear SVMs in linear time. In *Proceedings of the 12th ACM SIGKDD International Conference on Knowledge Discovery and Data Mining* (pp. 217–226).

Kang, W.-C., & McAuley, J. (2018). Self-attentive sequential recommendation. In *Proceedings of the 2018 IEEE International Conference on Data Mining (ICDM)* (pp. 197–206). https://doi.org/10.1109/ICDM.2018.00035

Kim, B., Jeong, Y., Lee, S., et al. (2024). *Large language models meet collaborative filtering: An efficient all-round LLM-based recommender system*. arXiv preprint arXiv:2404.11343.

Koren, Y., Bell, R., & Volinsky, C. (2009). Matrix factorization techniques for recommender systems. *Computer, 42*(8), 30–37.

McMahan, H. B., Holt, G., Sculley, D., Young, M., Ebner, D., Grady, J., Nie, L., Phillips, T., Davydov, E., Golovin, D., Chikkerur, S., Liu, D., Wattenberg, M., Hrafnkelsson, A. M., Boulos, T., & Kubica, J. (2013). Ad click prediction: A view from the trenches. In *Proceedings of the 19th ACM SIGKDD International Conference on Knowledge Discovery and Data Mining* (pp. 1222–1230). https://doi.org/10.1145/2487575.2488200.

References

Mehrotra, R., et al. (2018). Towards a fair marketplace: Counterfactual evaluation of recommender systems.

Mihalcea, R., & Tarau, P. (2004). TextRank: Bringing order into texts. In *Proceedings of EMNLP* (pp. 404–411).

Mikolov, T., Chen, K., Corrado, G., & Dean, J. (2013). Efficient estimation of word representations in vector space. In *Proceedings of ICLR*.

Pennington, J., Socher, R., & Manning, C. D. (2014). GloVe: Global vectors for word representation. In *Proceedings of EMNLP* (pp. 1532–1543).

Rendle, S. (2010). Factorization machines. In *2010 IEEE International Conference on Data Mining* (pp. 995–1000).

Rendle, S., Krichene, W., Zhang, L., & Anderson, J. (2020). *Neural collaborative filtering vs. matrix factorization revisited*. arXiv preprint arXiv:2005.09683.

Resnick, P., Iacovou, N., Suchak, M., Bergstrom, P., & Riedl, J. (1994). GroupLens: An open architecture for collaborative filtering of netnews. In *Proceedings of the 1994 ACM Conference on Computer Supported Cooperative Work* (pp. 175–186).

Sarwar, B., Karypis, G., Konstan, J., & Riedl, J. (2001). Item-based collaborative filtering recommendation algorithms. In *Proceedings of the 10th International Conference on World Wide Web* (pp. 285–295).

Shani, G., & Gunawardana, A. (2011). Evaluating recommendation systems. In *Recommender systems handbook* (pp. 257–297). Springer.

Sun, F., Liu, J., Wu, J., Pei, C., Lin, X., Ou, W., & Jiang, P. (2019). BERT4Rec: Sequential recommendation with bidirectional encoder representations from transformer. In *Proceedings of the 28th ACM International Conference on Information and Knowledge Management (CIKM)* (pp. 1441–1450).

Tan, J., Xu, S., Hua, W., Ge, Y., Li, Z., & Zhang, Y. (2024). *IDGenRec: LLM–RecSys alignment with textual ID learning*. arXiv preprint arXiv:2403.19021.

Wu, L., Zheng, Z., Qiu, Z., Wang, H., Gu, H., Shen, T., Qin, C., Zhu, C., Zhu, H., Liu, Q., Xiong, H., & Chen, E. (2023). *A survey on large language models for recommendation*. arXiv preprint arXiv:2305.19860.

Yi, X., Yang, J., Hong, L., et al. (2019). Sampling-bias-corrected neural modeling for large corpus item recommendations. In *Proceedings of the 13th ACM Conference on Recommender Systems* (pp. 269–277).

Ziegler, C.-N., McNee, S. M., Konstan, J. A., & Lausen, G. (2005). Improving recommendation lists through topic diversification. In *Proceedings of the 14th International Conference on World Wide Web* (pp. 22–32).

Zhang, J., Xie, R., Hou, Y., Zhao, W. X., Lin, L., & Wen, J.-R. (2023). *Recommendation as instruction following: A large language model empowered recommendation approach*. arXiv preprint arXiv:2305.07001.

Zhou, G., Zhu, X., Song, C., et al. (2018). Deep interest network for click-through rate prediction. In *Proceedings of the 24th ACM SIGKDD International Conference on Knowledge Discovery & Data Mining* (pp. 1059–1068).

Chapter 3
LLM-Enhanced Recommendation Systems

This chapter covers key LLM techniques that address traditional recommendation challenges to enhance existing systems. It begins with an overview of the importance of using LLMs to enhance recommendation systems, then covers key techniques for LLM enhancement, including tokenization, embeddings for richer data representation, and ANN algorithms for efficient retrieval. We'll also explore how LLMs assist in data labeling and evaluation, enabling more accurate and scalable recommendation systems. We then close the chapter with two tutorial examples, one demonstrates the use of LLMs for topic classification and item similarity labeling, and another shows how to combine LLM embeddings with traditional ranking models for news recommendation.

3.1 Overview

In the previous chapter, we discussed traditional recommendation systems, including collaborative filtering and content-based models. While effective, these systems face persistent challenges such as data sparsity and the cold-start problem. Large Language Models (LLMs) offer new capabilities to address these issues by introducing context-aware reasoning, cross-modal understanding, and language-driven personalization.

However, LLMs also come with significant limitations when deployed as stand-alone recommender systems. Traditional systems are engineered for real-time interaction data (e.g., clicks, impressions) with low-latency, high-throughput serving requirements. LLMs, in contrast, are computationally intensive, slower at inference time, and typically require batching or preprocessing pipelines to operate efficiently at scale. Furthermore, traditional recommenders are tightly optimized for domain-specific objectives like click-through rate (CTR), conversion rate, or revenue maximization, whereas LLMs prioritize general semantic reasoning rather than task-specific optimization.

These limitations highlight why LLMs are well positioned as *enhancers* rather than *standalone recommenders*. This is similar to their role in modern search engines, where models like BERT improve query understanding, document ranking, and personalization. In the same spirit, LLMs can enrich recommender systems by improving user modeling, contextual understanding, and item representation. For instance, LLMs can generate embeddings that capture item semantics across modalities (text, images), infer latent user preferences from behavioral or textual cues, and generate personalized recommendations even in cold-start settings.

This hybrid paradigm of combining LLMs with traditional pipelines offers a promising path forward. It leverages the precision and scalability of traditional recommenders while extending their expressiveness and adaptability through language-based reasoning. The result is a new class of recommendation systems that are more personalized, explainable, and robust to sparse data environments.

3.2 LLM Tokenization for Recommendations

Traditional recommendation systems rely on predefined features (e.g., genres, user IDs) and explicit interactions (e.g., clicks, ratings). However, they face several key limitations:

1. **Limited Nuance**: Predefined features fail to capture the full semantic range of item attributes.
   - **LLM Tokenization**: LLMs generate semantic tokens (e.g., "lightweight," "sturdy") that capture nuanced user intent and item characteristics, enabling more refined recommendations.

2. **Cold-Start and Sparse Data:** New users/items and low interaction density hinder recommendation quality.
   - **LLM Tokenization:** Represents users and items through semantic themes (e.g., "vegan leather bag"), enabling more effective recommendations without extensive history.

3. **Contextual Blindness**: Traditional systems struggle with interpreting contextual signals or unstructured data.
   - **LLM Tokenization**: LLMs capture implicit preferences and context (e.g., sentiment in reviews, user intent), enhancing recommendation accuracy.

LLM tokenization transforms raw data into semantically rich tokens, improving understanding, mitigating cold-start issues, and enabling more personalized, context-aware recommendations.

3.2 LLM Tokenization for Recommendations

3.2.1 LLM Tokenization Workflow

LLM tokenization transforms diverse data types—such as text, categorical variables, numerical values, and multi-modal content—into discrete tokens that preserve semantic meaning and enable sequence modeling. For example, in modeling user-item interactions, we can construct tokenized sequences like:

- **User ID**: Encoded as a token, e.g., USER_123
- **Item ID**: Encoded as ITEM_456
- **Interaction Type**: Tokens like CLICK, PURCHASE
- **Timestamp**: Discretized into temporal tokens like TIME_MORNING, TIME_EVENING

These tokens form a unified sequence:

[USER_123, CLICK, ITEM_456, TIME_EVENING]

This setup allows LLMs to learn patterns from sequential user behavior, capturing both temporal and interactional dynamics.

In addition to raw identifiers, LLMs can enhance tokenization by appending *semantic descriptors* to user and item tokens. For instance:

- USER_123 (budget-conscious)
- ITEM_456 (sci-fi, top-rated, under $20)

These augmented tokens help LLMs generate richer embeddings and interpretable features for downstream tasks (Tan et al., 2024; Geng et al., 2022). In content-based filtering, such tokens improve personalization by aligning user preferences with item properties like "eco-conscious" or "family-friendly."

3.2.1.1 Text Data

Steps

1. **Normalization**: Clean text (lowercase, remove special characters).
2. **Subword Tokenization**: Use pre-trained tokenizers (e.g., BERT's WordPiece, GPT's byte pair encoding).
   - Example: "durable laptop" → ["durable", "lap", "##top"].
3. **Semantic Enrichment**:
   - Apply domain-specific tokenization (e.g., "4K resolution" → ["4K", "resolution"]).

Use Case: Tokenizing product descriptions into themes (e.g., "gaming laptop," "budget-friendly").

3.2.1.2 Categorical Features

Approach

- **Embedding-Based Tokenization**:
  - Assign unique tokens to categories (e.g., USER_123, ITEM_456).
  - Map tokens to dense embeddings (e.g., USER_123 → 256-dim vector).
- **Hierarchical Tokenization**:
  - Group categories into hierarchies (e.g., electronics/laptops/gaming).

Use Case: Encoding user demographics (e.g., age_25–34, location_nyc).

3.2.1.3 Numerical Features

Methods

- **Discretization**: Bucketize values (e.g., price → PRICE_HIGH, PRICE_MEDIUM).
- **Text Conversion**: Represent numbers as text (e.g., "4.5 stars" → ["4.5", "stars"]).
- **Direct Embedding**: Use neural networks to embed raw values (e.g., time of day → 128-dim vector).

Use Case: Tokenizing user ratings (e.g., RATING_4.5).

3.2.1.4 Multi-modal Data

Strategy

1. **Unified Tokenization**:
   - Text: Subword tokens (e.g., BERT for product descriptions).
   - Images: Vision transformer tokens (e.g., ViT splits images into patch embeddings).
   - Metadata: Categorical/numerical tokens (e.g., BRAND_Apple).
2. **Cross-Modal Alignment**:
   - Map tokens to a shared space (e.g., CLIP aligns text and image tokens).

Example

- A product with text ("waterproof watch"), image (⌚), and metadata (PRICE_200) → tokens ["waterproof", "watch"], [IMG_EMBED], PRICE_200.

3.2.2 Integrating LLM Tokenization to Recommendation Systems

LLM-based tokenization goes beyond traditional text processing methods (e.g., TF-IDF, Bag-of-Words) by capturing semantic richness, context, and emerging vocabulary. These tokens form the foundation for enhanced user and item representation, enabling recommendation systems to reason over deeper meanings and dynamic trends.

3.2.2.1 Semantic Tokenization and Concept Extraction

LLMs tokenize text into subword units and semantically meaningful segments, allowing systems to capture fine-grained product and user attributes.

- **Context-Aware Tokenization**: LLMs segment inputs based on context, not just spelling. For instance, "eco-friendly" may be split into "eco" and "friendly," both tied to sustainability semantics (Devlin et al., 2019). This allows better grouping of items with shared environmental features.
- **Hierarchical Concept Extraction**: LLMs understand terms at different abstraction levels. A phrase like "long battery life" may be interpreted within higher order categories like "durability" or "electronics," offering more meaningful item clustering.

For example, a product description such as "4K OLED TV with immersive viewing experience" is tokenized into functional terms ("4K," "OLED") and experiential attributes ("immersive viewing"), which helps model both technical specs and user sentiment.

3.2.2.2 Hybrid Modeling with Semantic Tokens

Rather than relying solely on ID-based representations, LLM-generated tokens can bridge traditional and deep learning approaches.

- **Textual Identifiers for IDs**: Replace raw item/user IDs with semantically meaningful token strings like "WirelessNoiseCancellingHeadphones" instead of "Item_123." These improve generalization across similar items and users (Tan et al., 2024).
- **Cross-Modal Alignment**: By mapping textual tokens (e.g., "sunset view") to embeddings in other modalities (e.g., image features), LLM token spaces support multi-modal recommendation tasks (Radford et al., 2021).

3.2.2.3 Dynamic Trend Adaptation

LLMs adapt to language evolution and domain shifts without manual intervention.

- **Emerging Lexicon Support**: Newly coined terms like "phygital" or "deinfluencing" are naturally tokenized using subword priors learned from large corpora. This allows models to stay current with evolving language.
- **Domain-Aware Disambiguation**: LLMs use attention to differentiate between meanings depending on context. For example, "viral" in a medical setting vs. social media is disambiguated automatically during tokenization.

3.3 Embeddings from Unstructured Data

LLM embeddings offer several advantages over traditional feature representations in recommendation systems:

- **Pre-trained Knowledge and Ease of Use**: LLMs provide embeddings without requiring custom training, leveraging their vast pre-trained knowledge to capture nuanced relationships (e.g., linking "yoga mat" and "fitness tracker" via wellness themes). This reduces development time and cost while ensuring high-quality representations.
- **Context-Aware and Cold-Start Resilience**: LLM embeddings adapt to context (e.g., disambiguating "bank" in financial vs. geographic settings) and support dynamic personalization by integrating signals such as reviews, clicks, or user bios. They also mitigate cold-start issues by generating meaningful representations from item metadata or initial user inputs.
- **Cross-Modality Alignment**: LLMs project text, images, and other modalities into a unified embedding space (Radford et al., 2021). This enables seamless cross-modal retrieval (e.g., text queries → image results) and joint understanding of multi-modal item attributes (e.g., product images + reviews). For example, a query for "minimalist Scandinavian furniture" retrieves both product images and matching textual descriptions.

3.3.1 Obtaining LLM Embeddings

1. **Textual Data**:
    - Use pre-trained LLMs (e.g., GPT, BERT) to generate embeddings.
    - Example:

3.3 Embeddings from Unstructured Data   77

```
from transformers import GPT2Tokenizer, GPT2Model
tokenizer = GPT2Tokenizer.from_pretrained('gpt2')
model = GPT2Model.from_pretrained('gpt2')
inputs = tokenizer("durable running shoes for marathon
training", return_tensors="pt")
outputs = model(**inputs)
embeddings = outputs.last_hidden_state.mean(dim=1)   #
Pooling to get a single vector
```

2. **Non-text Data (Images, Videos)**:
   - Convert non-text data to text using generative models (e.g., image captioning with BLIP).
   - Align cross-modal data using models like CLIP:

```
import torch
from transformers import CLIPProcessor, CLIPModel
model = CLIPModel.from_pretrained("openai/
clip-vit-base-patch32")
processor = CLIPProcessor.from_pretrained("openai/
clip-vit-base-patch32")
image = Image.open("sneakers.jpg")
inputs = processor(text=["red sneakers"], images=image,
return_tensors="pt", padding=True)
outputs = model(**inputs)
image_embeddings = outputs.image_embeds
text_embeddings = outputs.text_embeds
```

3.3.2 Storing Embeddings

In Sect. 1.4.2, we have introduced the use of vector database or retrieval packages for the storage and retrieval of embeddings.

- **Vector Databases or Retrieval Packages**: Use specialized databases like Pinecone, Weaviate, or FAISS for efficient storage and retrieval. Example with FAISS:

```
import faiss
dimension = 768   # Embedding dimension
index = faiss.IndexFlatL2(dimension)
index.add(embeddings.numpy())   # Add embeddings to the index
```

- **Metadata Association**: Store metadata (e.g., user IDs, item descriptions) alongside embeddings for interpretability and filtering.

3.3.3 Evaluating Embeddings

Embeddings play a crucial role in modern recommendation systems and LLM-based retrieval, encoding semantic relationships between users, items, and queries. To ensure their effectiveness, we evaluate embeddings using the following methods.

3.3.3.1 Retrieval Quality

Measures how well embeddings retrieve relevant items from a candidate pool. Common metrics include:

- **Recall@k**: Proportion of relevant items found in the top-k recommendations.
- **Precision@k**: Fraction of top-k retrieved items that are relevant.

Example (E-Commerce Recommendations)
- Suppose a user searches for "durable running shoes."
- The system retrieves embeddings for shoes like *Nike Pegasus*, *Adidas Ultraboost*, and *Hoka Clifton*.
- If only *Ultraboost* and *Pegasus* are truly durable (based on product specs), and the system retrieves them in the top-5, then:
    - Recall@5 = 2/2 = 100% (all relevant items retrieved).
    - Precision@5 = 2/5 = 40% (only 2 of 5 recommendations are correct).

3.3.3.2 Labeled Similarity Data

Evaluates whether embeddings align with human-judged similarity. Benchmarks like STS-B (Semantic Textual Similarity Benchmark) provide labeled pairs with similarity scores (0–5).

Example (Text Embeddings in Recommendations)
- Compute cosine similarity between embeddings of:
    - "The Godfather" and "Goodfellas" (human score: 4.5/5—both crime dramas).
    - "The Godfather" and "Toy Story" (human score: 0.5/5—dissimilar genres).
- A good embedding model should reflect this with high cosine similarity (~0.8) for the first pair and low (~0.1) for the second.

Methodology
- Calculate *Spearman's rank correlation* between embedding similarities and human scores.
- Strong correlation (>0.6) indicates the embeddings capture semantic relationships well.

3.3.3.3 Downstream Task Performance

Embeddings should improve performance in real-world tasks like *CTR prediction* or *ranking*.

Example (News Recommendation)
- Train two models:
  1. **Baseline**: Uses one-hot encoded article IDs.
  2. **Embedding-Based**: Uses article title embeddings (e.g., from LLMs like BERT).
- Compare their AUC-ROC in predicting user clicks.
- If the embedding model improves AUC by 5%, it confirms better generalization.

3.4 LLM-Augmented Retrieval

In Chap. 1, we introduced the fundamentals of retrieval in recommendation systems, distinguishing between sparse retrieval (e.g., TF-IDF, BM25) and dense retrieval (e.g., embedding-based methods). In Sect. 3.3, we further explored how user and item representations can be enriched through LLM-generated embeddings.

This section focuses on retrieval mechanisms, specifically how to efficiently retrieve relevant items given user or item embeddings. We will first examine dense retrieval algorithms such as Locality-Sensitive Hashing (LSH), ANNOY, and Hierarchical Navigable Small World Graphs (HNSW), followed by a discussion of industrial-grade tools like FAISS and SCANN that implement these algorithms at scale.

Finally, we explore how Large Language Models (LLMs) can enhance retrieval systems beyond embedding generation. We focus on three practical methods: query rewriting, contextual augmentation, and hybrid retrieval, showing how LLMs contribute semantic understanding, contextual awareness, and flexibility to retrieval-based recommendation systems.

3.4.1 Dense Retrieval

Exact nearest neighbor (NN) search is computationally expensive, especially for large-scale datasets, as it requires comparing the query vector with every item in the dataset. To address this, Approximate Nearest Neighbor (ANN) algorithms are employed, trading off some accuracy for significantly faster retrieval.

ANN tools rely on similarity metrics to rank embeddings:

1. **Cosine Similarity**: Measures the angular distance between vectors, ideal for normalized embeddings.
2. **Dot Product**: Commonly used for ranking embeddings, especially in models like Two-Tower Networks.

Trade-offs in ANN

1. **Recall vs. Latency**: Higher recall often requires more computational resources, while lower latency may sacrifice some accuracy.
2. **Scalability**: ANN algorithms enable retrieval from millions of embeddings in real time, a critical requirement for modern recommendation systems.

ANN methods provide an effective balance between accuracy and computational efficiency, making them indispensable for large-scale recommendation systems.

3.4.1.1 Locality-Sensitive Hashing

Locality-Sensitive Hashing (LSH) maps high-dimensional data into lower dimensional space while preserving relative distances (Indyk & Motwani, 1998). Similar items are hashed into the same "bucket" with high probability. Figure 3.1 illustrates how the LSH algorithm uses random projections to hash data points onto lower

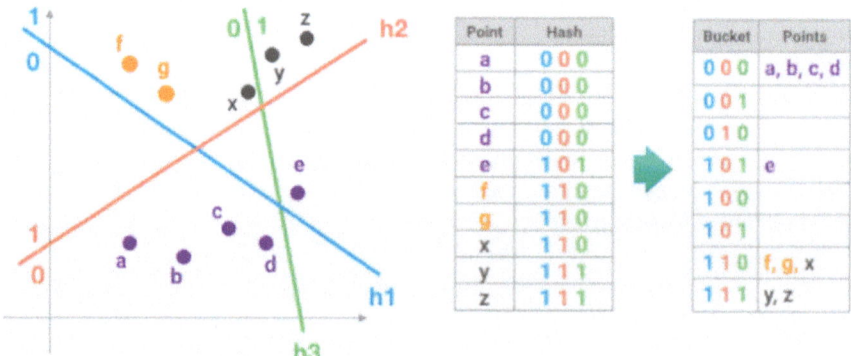

Fig. 3.1 LSH implementation using random hyperplanes (h_1, h_2, h_3) in 2D space

3.4 LLM-Augmented Retrieval

dimensional space. There are three separating hyperplanes, h_1, h_2, and h_3. We can take h_1 for example. h_1 separates data points *a*, *b*, *c* and *d* from the rest. We can assign *a*, *b*, *c* and *d* hash code 0, and the remaining points receive hashcode 1. In the same manner, we can assign hash codes based on h_2 and h_3. We then group the data points based on the three-digit hashcode.

Trade-off: High recall but requires large memory for hash tables.

3.4.1.2 Space-Partitioning Algorithms (e.g., KD-Trees, Annoy)

Space-partitioning algorithms accelerate nearest neighbor search by recursively dividing the embedding space into smaller subregions.

KD-Trees (Bentley, 1975)

Bentley introduced the *KD-tree*, which recursively splits the space using axis-aligned hyperplanes based on coordinate values (Bentley, 1975). Query vectors traverse the tree to reach leaf nodes containing candidate neighbors.

- *Efficient* for low-dimensional data (e.g., <20 dimensions).
- *Ineffective* in high dimensions due to the curse of dimensionality.

ANNOY (Spotify, 2015)

Spotify's ANNOY algorithm extends this idea using random hyperplane splits. Each tree partitions space until leaf nodes contain at most *k* items. At query time, multiple trees are traversed to collect and rank candidates.

- *Fast and scalable* for high-dimensional data.
- *Balances* speed and recall with lightweight indexing.

Figure 3.2 illustrates how ANNOY recursively partitions the embedding space into smaller subspaces, each holding *k* data points.

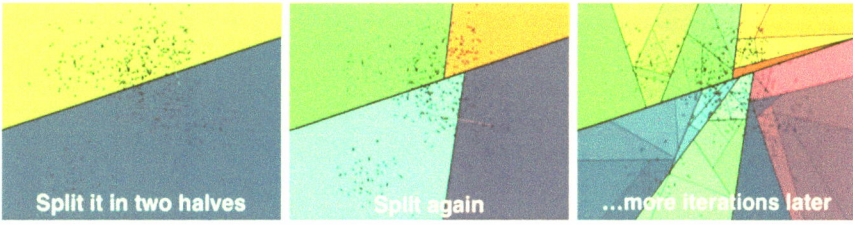

Fig. 3.2 Implementation of ANNOY algorithm through recursive partitioning

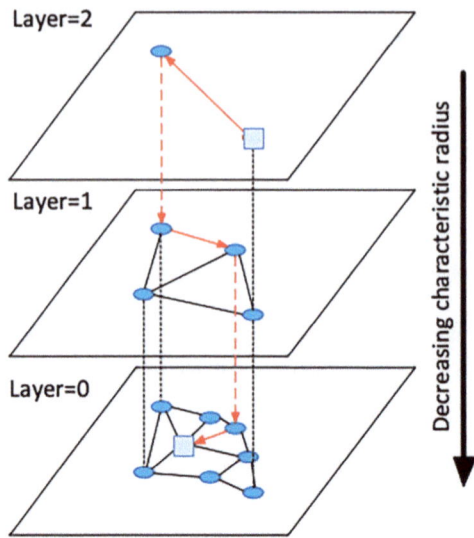

Fig. 3.3 Implementation of HNSW algorithm through hierarchical navigation

3.4.1.3 Graph-Based Traversal Algorithms (e.g., NSW, HNSW)

Navigable Small World (NSW) graphs represent embeddings as nodes, with edges linking similar items. During search, a greedy traversal moves from node to node based on distance, quickly converging on nearest neighbors.

Hierarchical NSW (HNSW) (Malkov & Yashunin, 2018) extends this by building multiple graph layers. Upper layers have fewer nodes and longer edges, enabling fast global navigation, while lower layers capture local precision.

Figure 3.3 demonstrates the HNSW process, where searches start at a high level and proceed layer by layer.

HNSW is often implemented in industrial tools like FAISS and ScaNN. It is commonly used in industry for large-scale search and recommendation systems due to its balance between performance and efficiency.

- **Hierarchical Layers**: Data points are assigned to different levels, with coarser connections at higher levels and finer connections at lower levels.
- **Greedy Search**: Queries start at the highest level and navigate downwards to find approximate nearest neighbors.

Trade-Offs

- **Accuracy**: HNSW achieves near-exact nearest neighbor search with high recall.
- **Memory Overhead**: The hierarchical structure requires additional memory but is still more efficient than LSH.
- **Parameter Tuning**: Requires tuning of key parameters—M (max neighbors per node) and efConstruction (search depth during graph building)—to balance accuracy, speed, and memory use.

3.4.2 Industrial Tools for Dense Retrieval

Dense retrieval at scale requires not only accurate algorithms but also efficient and production-ready tooling. Industrial-grade libraries such as FAISS and ScaNN extend foundational Approximate Nearest Neighbor (ANN) techniques to support fast, scalable retrieval in real-world recommendation and search systems.

3.4.2.1 FAISS (Facebook AI Similarity Search)

FAISS is a widely used library optimized for high-performance similarity search, particularly at large scales (Johnson et al., 2019). It supports:

- **Inverted File Index (IVF):** Clusters embeddings to reduce search space.
- **Product Quantization (PQ):** Compresses vectors to reduce memory usage.
- **GPU Acceleration:** Enables fast retrieval over millions of embeddings.

Applications

- Real-time recommendations (e.g., Netflix, e-commerce search).

Trade-Offs

- High recall with tunable latency.
- PQ reduces memory at the cost of slight accuracy drop.

3.4.2.2 ScaNN (Scalable Nearest Neighbors by Google)

ScaNN (Guo et al., 2020) balances accuracy and efficiency using a hybrid of quantization and search refinement:

- **Anisotropic Vector Quantization**: Enhances recall by considering vector directions.
- **Reordering Step**: Refines top results with exact distance computations.

Applications

- Large-scale image/text retrieval and recommendation systems.

Trade-Offs

- Highly scalable with low latency.
- Requires tuning for optimal recall-cost trade-off.

3.4.3 *LLM-Enhanced Retrieval*

LLMs can enrich retrieval pipelines by improving how queries are formulated and adding contextual signals.

3.4.3.1 Query Rewriting

What It Is: Query rewriting refers to the process of reformulating user queries to improve their alignment with indexed content in a retrieval system. This is especially valuable when user input is vague, short, or semantically underspecified.

How LLM Enhances: Large Language Models (LLMs) enhance query rewriting by paraphrasing, expanding, or contextualizing the original query using pre-trained semantic knowledge.

Example:

```
original_query = "affordable headphones"
prompt = f"Rewrite this query to include related features
and synonyms: '{original_query}'"
# LLM output: "budget-friendly wireless over-ear headphones
with good battery life"
```

Impact: This expansion captures semantically related terms that may not be explicitly present in the original query, improving both precision and recall in retrieval tasks.

3.4.3.2 Contextual Augmentation

What It Is: Incorporating additional context (e.g., user history, session data) into the query vector before retrieval (Zuo et al., 2022; Anand et al., 2023).

How LLM Enhances: LLMs can summarize or augment the user context into a richer query prompt or embedding that captures latent preferences.

Example:

```
user_history = ["User purchased a yoga mat and
resistance bands"]
prompt = f"Generate a contextual query for recommending
fitness products based on: {user_history}"
# LLM output: "Home workout gear for strength and flexibility training"
```

Impact: Boosts personalization by tailoring the query to implicit user needs.

3.5 LLM-Based Data Labeling and Evaluation

Labeling training data and evaluating recommendation quality are resource-intensive tasks traditionally requiring significant human effort (e.g., Mechanical Turk) or costly infrastructure. LLMs offer a *cost-effective*, *scalable*, and *ready-to-integrate* solution by:

- *Generating synthetic data* or *pseudo-labels* to reduce reliance on platforms like Mechanical Turk.
- *Enabling zero-shot labeling* (e.g., classifying product reviews as "positive/negative" without fine-tuning).
- *Integrating directly into production pipelines* for real-time label refinement.

3.5.1 LLM-as-a-Judge for Recommendation Evaluation

Large Language Models (LLMs) are increasingly used as evaluation judges to assess the relevance of recommended content, especially when traditional metrics fall short of capturing semantic nuance or user intent (Liu et al., 2023). Instead of relying solely on click-through data or human-labeled samples, LLMs provide flexible, context-aware evaluations through natural language reasoning.

3.5.1.1 Key Frameworks

- **Zeng et al.** (2023): Introduced LLM-as-a-judge benchmarks for fairness and accuracy in recommendations.
- **Liu et al.** (2023): Demonstrated LLM judges outperform human annotators in consistency.

3.5.1.2 General Workflow

1. **User History Representation**: Summarize the user's past interactions (e.g., purchases, ratings) as a textual prompt.
   - Example: "User has purchased running shoes and gym accessories, and values affordability."
2. **Item Representation**: Describe the recommended item in a similar textual format.
   - Example: "Fitness tracker that tracks heart rate, steps, and calories burned. Affordable with positive reviews."
3. **Relevance Query**: Ask the LLM to assess the item's relevance based on the user's profile.

- Example query: "Is this fitness tracker relevant to a user interested in improving fitness and valuing affordability?"
4. **LLM Response**: The LLM generates a relevance score or natural language response.
   - Example response: "Yes, this fitness tracker aligns with the user's interests and budget."
5. **Metric Calculation**: Use metrics like precision, recall, F1-score, or NDCG to evaluate the system's performance.

3.5.1.3 Two Approaches

Generative Evaluation (Direct Scoring)

In the generative approach, LLMs directly generate relevance scores or pseudo-labels for content based on user history. For example:

- Given a user's search history, the LLM generates product pairs and assigns scores, which are then used to train downstream recommenders.
- LLMs can also create *pseudo queries* or *pseudo documents* from real content, expanding the training dataset with diverse query-document pairs.

Wang et al. (2023) introduced a "generate-then-filter" pipeline for training data creation:

- **Generate**: Create synthetic queries or document pairs aligned with user behavior (e.g., new item + past user preferences).
- **Filter**: Score generated samples with a ranking LLM (e.g., monoT5, GPT-4).
- **Train**: Use top-scoring samples to fine-tune retrieval or ranking models.

Discriminative Evaluation (Relative Judgments)

In the discriminative approach, LLMs assess the relative relevance of content rather than generating standalone scores. As illustrated with a search retrieval example in Fig. 3.4, this approach includes three methods:

- **Pointwise**: LLMs evaluate the relevance of a single document to a user (e.g. "Is this document relevant?").
- **Pairwise**: LLMs compare two documents to determine which is more relevant to a user.
- **Listwise**: LLMs rank a list of documents based on relevance to a user.

This method enables fine-grained control over relevance ranking and can be used to generate training signals or rerank outputs. We can adopt the same framework for recommendation evaluation.

3.5 LLM-Based Data Labeling and Evaluation

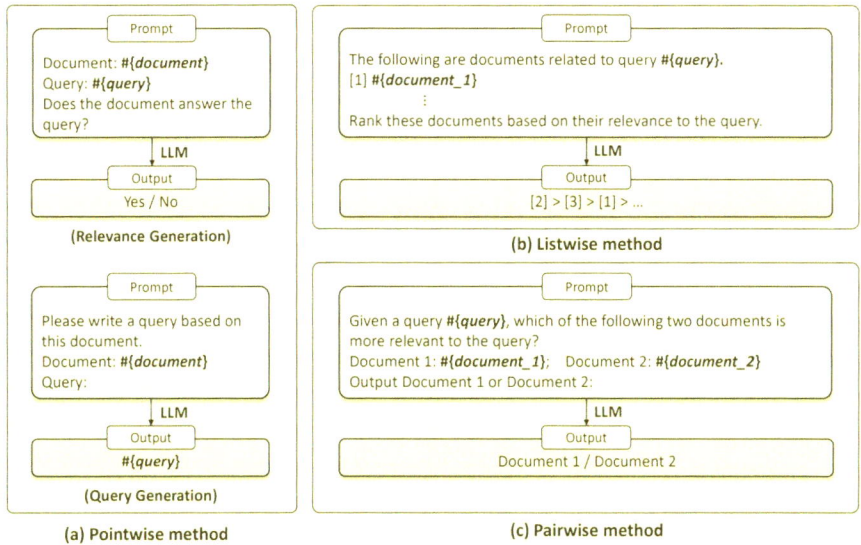

Fig. 3.4 (a–c) Pointwise, pairwise, and listwise methods for relevance scoring

3.5.2 Human-Assisted LLM Labeling

Labeling high-quality data has traditionally relied on manual efforts through platforms like Mechanical Turk, often resulting in high costs, inconsistent quality, and slow turnaround. Advances in Large Language Models (LLMs) have transformed this process, enabling scalable and semi-automated labeling pipelines that combine the strengths of AI with human oversight.

Many data labeling platforms now integrate LLMs to pre-label data, with human annotators verifying or correcting the outputs. This hybrid model improves accuracy while drastically reducing labeling time and cost. Examples include:

- Labelbox and Amazon SageMaker Ground Truth: Use AI to pre-annotate data, then route low-confidence examples to human reviewers.
- Scale AI and SuperAnnotate: Offer HITL frameworks for structured data labeling, combining LLMs with task-specific human validators.
- CVAT (Computer Vision Annotation Tool): While focused on visual tasks, it supports interactive refinement of AI-generated segmentations.

This approach reduces human burden while retaining the precision needed for critical applications like recommendation system evaluation or fine-grained sentiment labeling.

Example Task Label 1M product images for "similar item" recommendations in e-commerce. The LLM-assisted labeling pipeline works as below:

1. **Pre-labeling**: Use CLIP (Radford et al., 2021) to embed images and group them into semantic clusters (e.g., "high heels," "running shoes"). Radford et al. (2021)

showed CLIP achieves 75% zero-shot accuracy on product categorization tasks, reducing initial labeling effort by 50%.
2. **Human Review**: Annotators verify/reject clusters (e.g., correct "stilettos" vs. "sneakers" misgroupings). Similar to the HITL framework in *SageMaker Ground Truth* (Amazon, 2022), where human review of AI pre-labels cut errors by 40% compared to pure automation.
3. **Active Learning**: Misclassified samples fine-tune a GPT-4V (OpenAI, 2023) model to improve granularity (e.g., distinguishing "trail running" vs. "road running" shoes). Wu et al. (2023) demonstrated that active learning with LLMs reduces labeling costs by 58–63% while maintaining 96–98% accuracy in fashion recommendation systems.

This pipeline allows for rapid generation of millions of labeled examples, while preserving quality through targeted human oversight. By merging LLMs with interactive tools and HITL platforms, teams can build more accurate and scalable labeling workflows—essential for training, evaluation, and refinement of modern recommendation systems.

3.6 Tutorial: Topic Classification and Item Similarity Labeling Using LLMs

3.6.1 Overview

This tutorial demonstrates how Large Language Models (LLMs) can be applied to two key content understanding tasks: *topic classification* and *item similarity labeling*. These tasks are essential for organizing and analyzing diverse types of content—such as news articles, product descriptions, and short-form videos—where it is often necessary to assign content to appropriate categories and identify meaningful similarities between item pairs.

Goal of the Tutorial
1. *Understand* how to use LLMs for topic classification and item similarity labeling.
2. *Learn* best practices for designing prompts, processing large datasets, and evaluating labeling quality.
3. *Apply* these techniques to real-world datasets, such as news articles, to derive actionable insights.

3.6.2 Experimental Design

The study is designed to evaluate the effectiveness of LLMs in two labeling tasks:

1. **Topic Classification Labeling**: Classify news articles into predefined topics (e.g., politics, technology, sports, business, entertainment) using LLM-generated prompts.

2. **Item Similarity Labeling**: identify pairs of similar articles by combining embedding-based similarity filtering with prompt-based verification.

3.6.2.1 Dataset Setup

- A Kaggle dataset of BBC news articles is used, with each news article represented by its summary.
- The BBC news articles are organized in folders like "business," "tech," "entertainment," etc.; these folder names are considered ground truth labels for the news article.

3.6.2.2 LLM Choices

We consider two leading LLMs: DeepSeek-V3 (released by DeepSeek AI in December 2024) and GPT-4o-mini (released by OpenAI in July 2024).

3.6.2.3 Labeling Methods

- **Topic Classification**: The LLM is prompted to classify articles into predefined topics and return results in JSON format. We considered zero-shot prompting to begin with, and then added labeling guidelines after examining some hard cases.

```
Given the following article summary, classify it into
relevant Tier 1 topics from the list below.
Topics: ['business', 'tech', 'entertainment', 'sport',
'politics']
Return a JSON object with the format:
{
    "top_topics": ["topic1", "topic2", "topic3"],  // At
most 3 topics
    "primary_topic": "top_topic"  // Most relevant topic
}
Ensure the topics are chosen from the provided list.
Article Summary:
"Indonesia's government has confirmed it is considering
raising fuel prices by as
much as 30%. Indonesia pays subsidies to importers in order
to stabilise domestic
fuel prices, …."
```

- **Item Similarity**: Embeddings are generated using a pre-trained transformer model (all-MiniLM-L6-v2), and cosine similarity is computed to filter candidate pairs. Then LLM is prompted to verify if two articles are similar, with responses structured in JSON format for easy parsing.

```
You are a helpful assistant for text similarity analysis.
Are these two articles discussing the same topic?
Article 1: "{text1}"
Article 2: "{text2}"
Provide your reasoning and output in strict JSON format:
{{
 "reasoning": "Explain your decision briefly",
 "answer": "Yes" or "No"
}}
```

3.6.2.4 Evaluation Metrics

- **Topic Classification**: We evaluated the agreement between the two LLMs (Deepseek-V3 and GPT-4o-mini), and also compared LLM-generated labels with ground truth.
- **Item Similarity**: We used item similarity labeling as a demonstrative example. We manually analyze.

3.6.3 Results and Analysis

3.6.3.1 Topic Classification Labeling

```
=== GPT vs DeepSeek Agreement ===
Agreement Rate: 96.72%


=== Classification Accuracy ===
GPT Accuracy: 93.17%
DeepSeek Accuracy: 92.90%
=== GPT Confusion Matrix ===
primary_topic_gpt  business  entertainment  politics  sport  tech
category
business                443              0        56      8     3
entertainment             4            377         3      1     1
politics                  6              1       404      5     0
Sport                     1              0         0    510     0
tech                     14             31        11      6   339
```

3.6 Tutorial: Topic Classification and Item Similarity Labeling Using LLMs

```
=== DeepSeek Confusion Matrix ===
primary_topic_deepseek  business  entertainment  politics  sport  tech
category
business                     439              0        64      5     2
entertainment                  6            374         3      1     2
politics                      12              0       398      5     1
Sport                          0              0         0    511     0
tech                          19             23        12      2   345
```

1. **Agreement Rate**:

   - GPT-4o-mini and DeepSeek-V3 achieved a high agreement rate of 96.72% under a *zero-shot prompting* setup, indicating moderate task difficulty. While effective overall, further improvements can be made by analyzing disagreement cases and refining prompts with *labeling guidelines* and *few-shot examples* to enhance consistency and handle edge cases more reliably.

2. **Classification Accuracy**:

   - GPT-4o-mini: 93.17%, DeepSeek-V3: 92.90%.
   - Both models perform similarly, with GPT-4o-mini slightly outperforming DeepSeek-V3.

3. **Confusion Matrices**:

   - GPT-4o-mini:

     Strong performance in sport (510/510 correct) and entertainment (377/386 correct).
     Minor confusion between business and politics (classifying 56 "business" articles as "politics").

   - DeepSeek-V3:

     Excellent performance in sport (511/511 correct) and entertainment (374/386 correct).
     Slightly higher confusion between business and politics (classifying 64 "business" articles as "politics").

   Table 3.1 shows labeling results for three exemplar cases:

1. **German Music Crisis**: Both Ground Truth and GPT label this as "entertainment," which is accurate given the focus on the music industry, while Deepseek labels it as "business," possibly due to the mention of business models and industry decline.

Table 3.1 News articles summaries, their ground truth label and LLM labels by GPT and Deepseek

Article summary	Labels
German Music Crisis: Germany's music industry, once the third largest globally, is struggling due to piracy and outdated business models. Former Universal Music Germany head Tim Renner says it's like a "zombie," while others argue it's still successful in parts. The industry peaked in 1997 but has since declined amid digital disruption and private copying	{Ground truth: entertainment, GPT: entertainment, Deepseek: business}
Dortmund Financial Struggles: Borussia Dortmund, Germany's only stock-listed football club, warns of bankruptcy after posting record losses and missing stadium rent payments. Shares plummeted 23%, and experts say a €35M capital injection is needed. The club is under pressure to bring in external executives	{Ground truth: business, GPT: sport, Deepseek: business}
HP Ink Lawsuit: A US woman is suing HP, claiming their ink cartridges are programmed to expire. The lawsuit highlights rising frustration over high running costs of printers, despite falling purchase prices. HP uses chip technology to monitor ink levels, which critics say drives up user costs	{Ground truth: tech, GPT: business, Deepseek: tech}

2. **Dortmund Financial Struggles**: The Ground Truth and deepseek label this as "business," reflecting the financial issues faced by the football club. GPT tends to treat it as "sport," likely focusing on the football aspect.
3. **HP Ink Lawsuit**: The ground truth and deepseek label this as "tech," appropriate given the focus on technology and printer cartridges. GPT labels it as "business," possibly due to the mention of a lawsuit and costs.

3.6.3.2 Item Similarity Labeling

```
{
     "pair": [
      "The full Finance Bill, with the Budget measures in
it, would then be returned to the Commons after the
election, if Labour secures another term in office.If a May
election is called, there could be as little as 18 days
between the Budget and the announcement of a date for the
election.Tory shadow chancellor Oliver Letwin said:
\"We can be sure of two things: the Budget will contain
measures to attract votes, and it will not contain
the \u00a38
```

3.6 Tutorial: Topic Classification and Item Similarity Labeling Using LLMs 93

```
billion of tax rises which independent experts say are
inevitable if Labour wins the election.\"The Budget,
likely to be
the last before the General Election, will be at about 1230
GMT on that Wednesday, just after Prime Minister's
question time.The Tories say it is likely the Budget will
contain measures to attract votes.",
        "The full Finance Bill, with the Budget measures in
it, would then be returned to the Commons after the
election, if Labour secures another term in office.If a May
election is called, there could be as little as 18 days
between the Budget and the announcement of a date for the
election.The Budget, likely to be the last before the
General Election, will be at 1230 GMT on that Wednesday,
just after Prime Minister's question time.Chancellor
Gordon Brown will deliver his Budget to the House of Commons
on 16 March, the Treasury has announced."
    ],
    "similarity": 0.9054108262062073,
    "label": 1,
    "reasoning": "Both articles discuss the Finance Bill
and the Budget in the context of an upcoming General
Election, specifically mentioning the timing and implications
of the Budget. However, Article 1 focuses on the
political implications and reactions from the Tory shadow
chancellor, while Article 2 provides specific details about
the Chancellor delivering the Budget. Despite these
differences, the core topic of the Budget and its
relation to the
election is consistent across both articles."
    }
```

We monitor both the embedding similarity and binary labels indicating whether pairs of articles are similar. The scatter plot in Fig. 3.5 below illustrates the relationship between embedding similarity scores and binary labels. Notably, article pairs with a similarity score above 0.9 are consistently classified as similar by the large language model (LLM). The area under the curve (AUC) for predicting the binary outcome using the similarity score as a predictor is 0.7, indicating a moderate predictive capability.

Here are some interesting insights from this experiment:

1. **Efficiency Gains**: Combining embedding-based filtering with LLM-based verification reduces the number of API calls and improves scalability.
2. **Multi-label Classification**: For topic classification, allowing multiple labels per article could improve accuracy for ambiguous cases.
3. **Human Evaluation is Crucial**: Ambiguous cases require human intervention to ensure high-quality labels.
4. **Prompt Design Matters**: Clear and specific prompts improve the accuracy of both topic classification and similarity labeling.

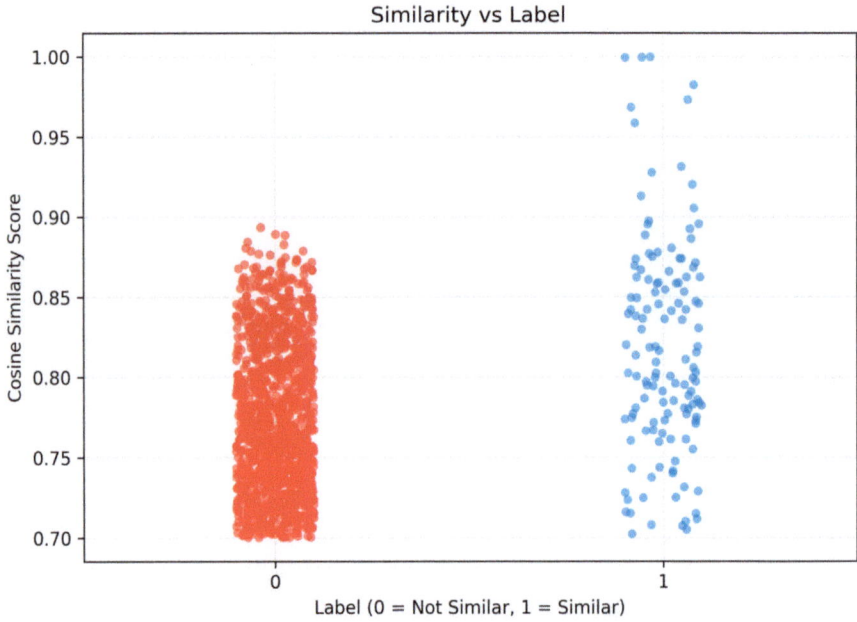

Fig. 3.5 Plot of cosine similarity score against binary label. Each point represents one pair of articles, with the vertical axis showing cosine similarity between article embeddings, and horizontal axis showing similarity labels obtained from LLM

3.6.4 Conclusions

This tutorial demonstrates the power of LLMs for automating labeling tasks in large datasets. By combining embedding-based techniques with LLM-based reasoning, we achieve scalable and accurate results. However, human evaluation remains essential for refining ambiguous cases and ensuring high-quality outputs. Future work could explore multi-label classification, fine-tuning LLMs for specific domains, and integrating feedback loops for continuous improvement.

3.7 Tutorial: News Recommendation by Combining Embedding with Learning-to-Rank Models

3.7.1 Overview

This tutorial demonstrates how to build a personalized recommendation system using user profiles and news articles. We show how to match users with relevant articles based on semantic similarity by leveraging pre-trained models to generate user and article embeddings. These embeddings are then used to retrieve candidate articles and train learning-to-rank models to produce personalized rankings.

3.7.1.1 Goal of the Tutorial

1. Learn how to *retrieve candidate articles* for each user using embedding-based semantic similarity.
2. Apply both *baseline similarity ranking* and *advanced learning-to-rank models* (e.g., LambdaMART) to sort the retrieved candidates.
3. Evaluate recommendation quality using standard metrics such as Precision@k, Recall@k, and NDCG@k.

The tutorial is designed for *beginners* and *advanced readers*, covering foundational concepts (e.g., embedding-based retrieval) and advanced techniques (e.g., LambdaMART ranking). We show a condensed version of this tutorial in the book text. The full version of the code is available at: https://github.com/qqwjq1981/springer-LLM-recommendation-system

3.7.2 Experimental Design

3.7.2.1 Data

- **User Profiles**: Generated using the Python package Faker, including fields like job_title, skills, hobbies, and summary.
- **News Articles**: Sourced from the BBC News dataset on Kaggle, containing title, summary, and category.

3.7.2.2 Retrieval Set Generation

- **Embeddings**: Use SentenceTransformer('all-MiniLM-L6-v2') to generate embeddings for user profiles and news articles.
- **Cosine Similarity**: Compute cosine similarity between user and article embeddings.
- **Retrieval**: For each user, retrieve the top-k most similar articles (e.g., top 50) as the candidate set.

3.7.2.3 Ground Truth Labeling

- **Prompt for Labeling**:

```
For each of the following user and article pairs, determine
the interest level.
Respond only with a single line per pair, using the
following format:
user_id, item_id, 1    ← for Interested
user_id, item_id, 0    ← for Not Interested
Do NOT add any explanations or additional formatting.
User ID: user_123
Profile: [User's profile summary here]
Article ID: article_456
Article: [Article summary text here]
```

- **Labeling**: Use the above prompt to simulate ground truth labels for user-item pairs.

3.7.2.4 Recommendation Approaches

- **Similarity-Based**: Use cosine similarity as the ranking score.
- **LambdaMART**: Train a ranking model using:
  - User and item embeddings.
  - Scalar cosine similarity as an additional feature.

3.7.2.5 Evaluation

- **Train-Test Split**: 80% of users for training, 20% for testing.
- **Metrics**: Precision@k, Recall@k, and NDCG@k for $k = 1, 5, 10$.

3.7.3 Results and Analysis

Table 3.2 summarizes evaluation metrics using similarity-based and LambdaMART-based ranking:

1. **Similarity-Based Approach**:
   - Achieves moderate performance, with Precision@10 of 0.42 and Recall@10 of 0.212.

2. **LambdaMART**:
   - Significantly outperforms the similarity-based approach across all metrics.
   - Achieves Precision@10 of 0.90 and Recall@10 of 0.541, demonstrating the effectiveness of learning-to-rank models.
   - Higher NDCG@10 (0.946) indicates better ranking quality.

3. **Key Insights**:
   - LambdaMART leverages both embeddings and cosine similarity as features, leading to more accurate recommendations.
   - The similarity-based approach is simpler but less effective, especially for top-k recommendations.

To extend this work, future iterations can incorporate:

- **User Behavioral History**: Enrich user modeling by integrating behavioral signals such as clicks, reading time, or past article interactions alongside static profiles.
- **Beyond Embedding Similarity in Retrieval**: Improve retrieval by leveraging hybrid methods, keyword-based search, or knowledge graph-enhanced retrieval instead of relying solely on embedding similarity.
- **Temporal Features**: Include temporal dynamics, such as article freshness or evolving user interests, to better capture time-sensitive relevance.
- **Deep Learning-Based Ranking Models**: Explore advanced ranking architectures (e.g., Transformers, attention-based networks) to model complex user-item relationships and improve ranking quality.

Table 3.2 Evaluation metrics by similarity-based ranking and LambdaMART

Metric	k	Similarity-based	LambdaMART
Precision@k	1	0.50	0.90
	5	0.44	0.88
	10	0.42	0.90
Recall@k	1	0.026	0.054
	5	0.113	0.270
	10	0.212	0.541
NDCG@k	1	–	–
	5	0.749	0.935
	10	0.767	0.946

3.7.4 Conclusion

This tutorial demonstrates how to build a personalized recommendation system using user profiles and news articles. By combining *embedding-based retrieval* with *advanced ranking models* like LambdaMART, we achieve significant improvements in recommendation quality. The results highlight the importance of leveraging both content-based features and learning-to-rank techniques for personalized recommendations.

References

Amazon Web Services. (2022). Scaling data annotation with SageMaker ground truth active learning. AWS Whitepaper.
Anand, A., Setty, V., Venkatesh, V., & Anand, A. (2023). *Context aware query rewriting for text rankers using LLM.* arXiv preprint arXiv:2308.16753.
Bentley, J. L. (1975). Multidimensional binary search trees used for associative searching. *Communications of the ACM, 18*(9), 509–517.
Devlin, J., et al. (2019). BERT: Pre-training of deep bidirectional transformers. In *ACL*.
Geng, S., Liu, S., Fu, Z., Ge, Y., & Zhang, Y. (2022). Recommendation as Language Processing (RLP): A Unified Pretrain, Personalized Prompt & Predict Paradigm (P5). In RecSys 2022- Proceedings of the 16th ACM Conference on Recommender Systems (pp. 299–315). (RecSys 2022 - Proceedings of the 16th ACM Conference on Recommender Systems). Association for Computing Machinery, Inc. https://doi.org/10.1145/3523227.3546767
Guo, R., Sun, P., Lindgren, E., et al. (2020). Accelerating large-scale inference with anisotropic vector quantization. In *Proceedings of the 37th International Conference on Machine Learning (ICML)* (pp. 3887–3896).
Indyk, P., & Motwani, R. (1998). Approximate nearest neighbors: Towards removing the curse of dimensionality. In *Proceedings of the 30th Annual ACM Symposium on Theory of Computing (STOC)* (pp. 604–613).
Johnson, J., Douze, M., & Jégou, H. (2019). Billion-scale similarity search with GPUs. *IEEE Transactions on Big Data, 7*(3), 535–547.
Liu, Y., et al. (2023). Judging LLM-as-a-Judge with MT-Bench. In *NeurIPS*.
Malkov, Y., & Yashunin, D. (2018). Efficient and robust approximate nearest neighbor search using hierarchical navigable small world graphs. *IEEE Transactions on Pattern Analysis and Machine Intelligence, 42*(4), 824–836.
OpenAI. (2023). GPT-4V technical report.
Radford, A., et al. (2021). Learning transferable visual models from natural language supervision. In *ICML*.
Spotify Engineering. (2015). Annoy: Approximate nearest neighbors in C++/Python.
Tan, H., et al. (2024). IDGenRec: LLM-RecSys alignment with textual ID learning. In *SIGIR*.
Wang, Y., et al. (2023). Synthetic data generation for recommender systems. In *ACM SIGIR*.
Wu, J., Zhang, Y., Chen, X., et al. (2023). Adaptive labeling for E-commerce recommendations with active learning.
Zeng, Z., Zhang, Y., Li, X., et al. (2023). Fairness-aware evaluation framework for LLM-based recommendation judges. In *Proceedings of the ACM Web Conference (WWW)*.
Zuo, S., Yin, Q., Jiang, H., Xi, S., Yin, B., Zhang, C., & Zhao, T. (2022). *Context-aware query rewriting for improving users' search experience on E-commerce websites.* arXiv preprint.

Chapter 4
LLM as Recommender

This chapter explores how Large Language Models (LLMs) can serve as end-to-end recommender systems. It covers key techniques such as prompting, fine-tuning, and cost optimization strategies including distillation, quantization, and caching. Practical design considerations are discussed to balance quality, cost, and scalability. The chapter concludes with two hands-on tutorials: one on fine-tuning LLMs for personalized movie recommendations, and another on applying knowledge distillation for efficient inference, offering practical insights for building LLM-based recommenders.

4.1 LLMs as Recommender End-to-End Workflow

This chapter builds on the foundation laid by Chap. 3, where we explored how LLMs can enhance traditional recommendation systems by addressing key challenges such as tokenization, embeddings, and data labeling. While LLMs as enhancers augment existing systems by adding semantic richness and handling complex queries, LLMs as recommenders attempt to directly generate recommendations from textual inputs.

Integrating large language models (LLMs) into recommendation systems involves a structured workflow that leverages their ability to process natural language and generate context-aware outputs. Figure 4.1 is a step-by-step guide to designing and implementing an LLM-based recommendation workflow, ensuring efficiency, relevance, and scalability.

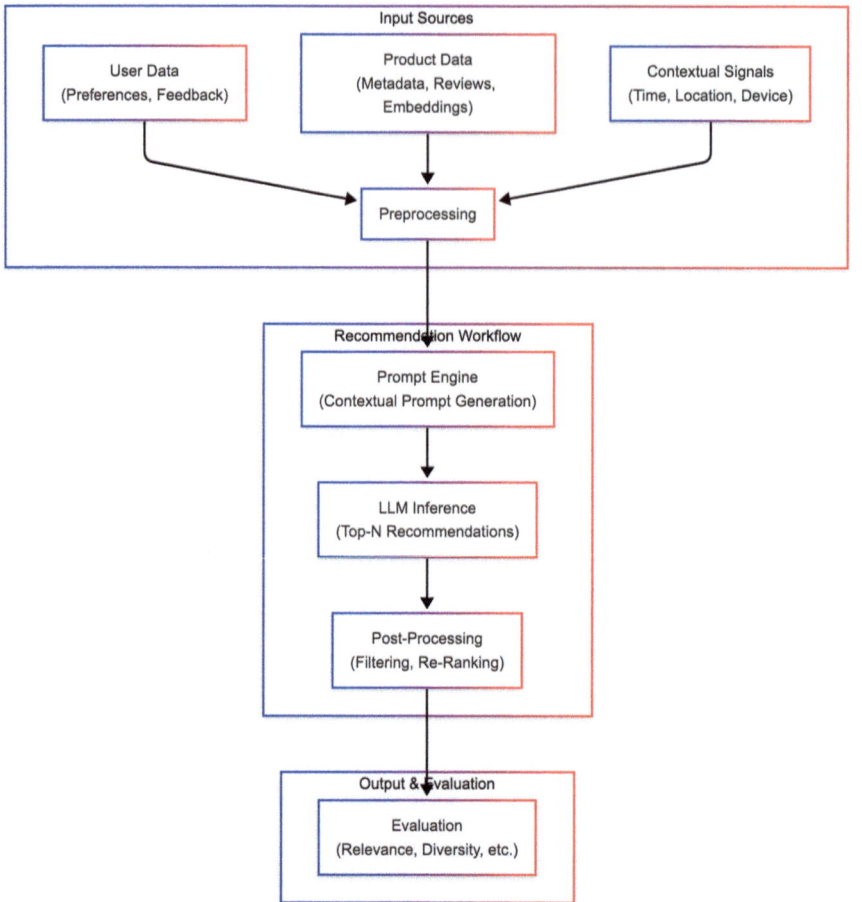

Fig. 4.1 System Workflow of LLM as Recommender

4.1.1 Step 1: Input Data Preparation

The foundation of recommendation systems lies in the quality and format of its input data. For LLM-based recommenders, data must be preprocessed into LLM-readable formats to maximize their interpretative capabilities:

- **Data Sources**:
    - **User Profile and History**: This includes static user profile and past interactions such as purchases, ratings, or browsing behavior. It can be represented in natural language (e.g., "User purchased hiking boots and rated outdoor gear highly") or structured JSON format.

4.1 LLMs as Recommender End-to-End Workflow

```
{
  "user_id": "user_12345",
  "profile": {
    "age": 32,
    "gender": "female",
    "location": "San Francisco, CA",
    "preferences": ["outdoor activities", "eco-friendly
products", "minimalist design"]
  },
  "history": {
    "purchases": [
      {
        "item_id": "item_987",
        "category": "hiking boots",
        "purchase_date": "2024-11-10",
        "price": 120.00
      }
    ],
    "ratings": [
      {
        "item_id": "item_987",
        "rating": 5,
        "review": "Very comfortable and durable for
long hikes."
      }
    ],
    "browsing": [
      {
        "timestamp": "2025-07-01T15:23:00Z",
        "category": "outdoor gear",
        "item_id": "item_321"
      }
    ]
  }
}
```

- **Item Data**: Descriptions of items (e.g., products, movies, or articles) should be simplified into key-value pairs or narrative formats (e.g., "A durable and lightweight sleeping bag for outdoor adventures").
- **Contextual Signals**: Additional context such as time, location, or events (e.g., "User is planning a weekend camping trip") can enrich recommendations.

- **Preprocessing**:
    - Summarize or structure user history into concise natural language descriptions.
    - Simplify item metadata into narratives or key-value pairs.
    - Natural language representations reduce preprocessing complexity and leverage LLMs' ability to interpret unstructured data effectively.

4.1.2 Step 2: Prompt Engine

The prompt acts as a structured interface between input data and the LLM, guiding the model to generate relevant and personalized recommendations. Well-crafted prompts translate user context, preferences, and constraints into a clear task for the LLM.

- **Instruction-Based Prompting**: Combine multiple tasks (e.g., candidate selection, ranking, and explanation) into a single prompt. For example:

```
User history: "Interested in fitness gear, previously
purchased running shoes and dumbbells."
Recommend five affordable items that align with these
interests, and explain each choice.
```

- **Minimal Optimization**:
  Ensure the prompt is concise and focused on the task. Avoid unnecessary complexity to prevent misinterpretation by the LLM.

4.1.3 Step 3: LLM Inference

Once the prompt is constructed, the LLM processes it to generate recommendations. This step involves interpreting the input and producing outputs that align with the user's needs.

- **Output Generation**:
  The LLM generates a ranked list of recommendations along with explanations, if required. Here we show top two recommendations as an example:

```
1. "Adjustable dumbbells - affordable and complements the
user's existing fitness gear."
2. "Yoga mat - suitable for fitness routines at home."
```

- **Key Considerations**:
  - Ensure the output is structured and easy to parse for downstream tasks.
  - Validate the LLM's reasoning to ensure recommendations are contextually appropriate.

4.1.4 Step 4: Post-Processing (Optional)

Post-processing refines the LLM's output to meet specific constraints or improve relevance.

- **Validation**:
  Check the output against constraints such as budget, availability, or other business rules. For example, remove items already purchased by the user.
- **Adjustments**:
  Perform minor edits to filter inappropriate or duplicate results. This step ensures the final recommendations are polished and user-ready.

4.1.5 Step 5: Evaluation

The evaluation step involves evaluating the workflow's performance and iterating to improve results.

- **Evaluation Metrics**:
  - **Relevance**: Do the recommendations align with the user's interests and context?
  - **Diversity**: Are the suggestions varied enough to avoid redundancy?
  - **Novelty**: Do the recommendations introduce new items distinct from the user's past interactions?
- **Iteration**:
  Refine prompts, incorporate user feedback, or fine-tune the LLM to address any shortcomings. Continuous iteration ensures the system evolves to meet user needs effectively.

4.2 Prompting for Recommendation

Prompting refers to structuring inputs for a large language model (LLM) to elicit specific and relevant responses. In the context of recommendation systems, prompting plays a central role in guiding the LLM to understand user preferences, interpret item descriptions, and generate tailored recommendations. The approach relies on leveraging the inherent knowledge and reasoning capabilities of LLMs without requiring additional training. Effective prompting bridges the gap between unstructured user behaviors and the structured outputs required for recommendation tasks. A comprehensive study by Xu et al. (2024) proposes a general framework for using LLMs in recommendation via prompt engineering, analyzing key aspects such as task formulation, user modeling, candidate item construction, and prompting strategies across various LLM types and recommendation scenarios.

For example, a prompt might describe a user's preferences (A user interested in high-tech gadgets and affordable options) and ask the LLM to suggest relevant items (Recommend three gadgets that fit these criteria). By embedding user context and task-specific instructions, prompting enables LLMs to generate personalized and relevant recommendations efficiently.

4.2.1 Prompting Techniques

4.2.1.1 Zero-Shot Prompting

Zero-shot prompting relies entirely on the LLM's pre-trained knowledge to perform a task without additional guidance or examples. The prompt is typically a direct query, and the model is expected to infer the appropriate format and content from the phrasing alone. This approach is especially suitable for open-ended recommendation queries where task intent is intuitive.

```
Prompt:
Recommend three laptops under $1,000 suitable for gaming.
Response:
Here are three options: Laptop A with a powerful GPU, Laptop
B with a high refresh rate display, and Laptop C with ample
storage for games.
```

4.2.1.2 Few-Shot Prompting

Few-shot prompting includes a few examples within the prompt to establish context, guiding the LLM to generate responses in line with desired patterns. This approach is effective when a specific recommendation style or logic needs to be demonstrated.

4.2 Prompting for Recommendation   105

```
Prompt:
Here is how to recommend movies based on genre preferences:
 User enjoys action films → Recommend "Mad Max" or
"Gladiator".
 User enjoys comedy films → Recommend "Superbad" or "The
Hangover".
 Now, recommend movies for a user who enjoys
romantic dramas.
Expected Model Output:
Recommend "The Notebook" or "Pride and Prejudice".
```

4.2.1.3 Instruction-Based Prompting

Instruction-based prompting involves telling the model explicitly what to do, often using imperative language or structured directions. While it may still involve zero-shot execution (no in-context examples), the task format, constraints, or expectations are directly described to guide the model's behavior.

```
Prompt:
List three budget-friendly smartphones that offer high-
quality cameras. Include brand names and one key feature
for each.
Response:
1. Xiaomi Redmi Note 12 - 50MP triple camera system
 2. Samsung Galaxy A54 - Optical image stabilization
 3. Google Pixel 6a - Excellent low-light performance
```

4.2.1.4 Chain-of-Thought (CoT) Prompting

CoT prompting encourages step-by-step reasoning, helping the LLM to break down complex tasks into logical steps. This is particularly useful for generating recommendations that require multifaceted analysis or prioritization.

```
Prompt: "Analyze the user's profile preferences. List
matching products in order of relevance. Then summarize the
top three."
Response: "The user prefers affordable fitness gear. Relevant
items are: 1) Budget fitness tracker, 2) Affordable running
shoes, 3) Discounted gym accessories. Top recommendations:
Fitness tracker for affordability, running shoes for utility,
and gym accessories for variety."
```

4.2.2 Prompting for Various Recommendation Tasks

Generative LLMs can be employed for various recommendation tasks, each with unique requirements. These tasks aim to suggest relevant items (movies, products, music, etc.) to users. Table 4.1 lists the breakdown of different tasks, their descriptions, and gives example prompts for LLM input.

Table 4.1 Example templates for various recommendation tasks

Task description	Example prompt
Rating prediction: Predicts a user's rating for an item they haven't interacted with yet	Here is the movie rating history of a user:"Guardians of the Galaxy": 9.2,"Transformers": 9.8.Based on the above rating history of the user, please rate a movie named "John Wick: Chapter 4" with a range of 1-10 points.
Top-K recommendation: Recommends a fixed number (k) of items most likely to interest the user	A user recently watched movies,"Avatar", "Godfather", "Forest Gump", "Lord of the Rings", "Matrix".Based on the watch history, please recommend 5 candidate movies that the user might be interested in from the following list.
Conversational recommendation: Incorporates dialogue with the user to refine recommendations based on feedback	Pretend you are a movie recommender system. I will give you a conversation between a user and you (the recommender). Based on the conversation, reply with 20 movie recommendations only--no explanations or extra sentences. Here is the conversation:User: I really enjoyed Interstellar and Inception.Recommender: Got it! You like sci-fi with strong narratives and visuals. Any genres you don't like?User: I'm not into horror or overly romantic stuff. Recommender: Noted. What about animated films or thrillers? User: Thrillers are great. Not a huge fan of animated films though.
Explanation generation: Provides explanations for why certain items are recommended	A new movie named "The Godfather Part II" is recommended to a user, who has recently watched movies: "12 Angry Men", "Goldfinger", "Casino Royale". Please explain the reasons.
Sequential recommendation: Takes into account the order of user interactions to suggest the next item	Alice, who enjoys comedy and action movies and recently watched "The Matrix", is looking for a new movie recommendation. Based on her preferences and the following movie descriptions, recommend the movie that best suits her taste: "The Raid 2: Berandal" (Action, Crime), "21 Jump Street" (Comedy, Action), "The Lord of the Rings: The Fellowship of the Ring".

4.2.3 Prompt Design Practical Tips

Well-structured prompts are essential for guiding LLMs to produce accurate, relevant, and context-aware recommendations. Poorly designed prompts may lead to ambiguous or irrelevant responses, undermining the system's utility. Here are some common practices in prompt design:

1. **Incorporating User Context**: Include detailed information about user preferences, history, or goals to enhance personalization.
   - *Example:* "User has purchased hiking boots and camping gear. Recommend three items for outdoor adventures."
2. **Balancing Detail and Conciseness**: Ensure the prompt is informative without overwhelming the model's context window. Use summaries or selective inclusion of data.
   - *Example:* Summarize a long purchase history into key preferences like "favors eco-friendly and budget-conscious products."
3. **Iterative Refinement**: Test and refine prompts iteratively to improve response quality. Analyze LLM outputs to identify ambiguities or errors and adjust accordingly.
4. **Handling Context Length Limits**: LLMs have finite context windows. Employ strategies like prioritizing recent or relevant interactions and summarizing older data.
   - *Example:* For a user with extensive history, focus on recent purchases related to the current query.
5. **Clarity and Specificity**: Prompts should provide clear instructions and sufficient context to avoid vague responses.
   - *Example:* Instead of "Recommend a product," use "Recommend an affordable fitness tracker for a user interested in outdoor activities."

By employing thoughtful prompt design, LLM-based recommendation systems can harness the full potential of these models to deliver contextually rich and highly personalized recommendations.

4.3 Fine-Tuning LLMs for Recommendation

While prompting provides a lightweight and flexible way to leverage large language models (LLMs) for recommendation, it often lacks the task-specific grounding needed for high-stakes or domain-sensitive scenarios. To achieve stronger performance, especially in specialized domains, additional investment in *pre-training and fine-tuning* is required. These approaches demand more resources and data curation but offer significantly improved accuracy, adaptability, and personalization.

One of the most promising directions is *pre-training LLMs specifically for recommendation tasks.* Instead of relying solely on general-purpose language knowledge, these models are trained from the ground up using recommendation-centric objectives and data formats. A notable example is P5 (Geng et al., 2023). P5 frames various recommendation tasks such as next-item prediction, review generation, and explanation into text-to-text problems. This unified formulation enables LLMs to handle a broad range of use cases while leveraging the expressiveness of natural language.

To support systematic pre-training and evaluation, OpenP5 (Geng et al., 2023) introduces a standardized benchmark that spans multiple recommendation paradigms, including sequential, knowledge-aware, and multi-modal tasks. OpenP5 provides curated datasets and prompt templates that align with real-world user-item interactions, serving as a foundation for training and evaluating general-purpose or domain-specific recommendation LLMs.

Building on pre-trained models, *fine-tuning* is essential to adapt LLMs to downstream recommendation tasks or domains. This step refines the model's parameters using supervised data tailored to specific contexts, such as product categories, user segments, or regional markets. A particularly efficient variant is Low-Rank Adaptation (LoRA) (Hu et al., 2021), which injects trainable low-rank matrices into each transformer layer while freezing the majority of the model weights. This reduces computational costs and facilitates frequent updates, making it ideal for real-world recommendation platforms that require responsiveness to new trends or user behaviors. For recommendation-specific fine-tuning, Wu et al. (2023) provide a comprehensive review of strategies designed to adapt general-purpose LLMs to recommender system tasks. In the following section, we explore three key fine-tuning strategies:

1. Fine-tuning for recommendation task formats
2. Fine-tuning for domain-specific knowledge
3. Fine-tuning for capturing user-level personalization

4.3.1 Instruction Fine-Tuning for Recommendation

Instruction fine-tuning trains LLMs to interpret natural language instructions and generate personalized recommendations. Research work has shown that instruction-tuned models significantly outperform standard LLMs on task-oriented recommendation tasks by better aligning model outputs with user intent. Notably, Zhang et al. (2023) proposed viewing recommendation as an instruction-following problem and demonstrate that a fine-tuned open-source LLM (Flan-T5-XL) can surpass even GPT-3.5 on multiple recommendation benchmarks, highlighting the value of structured instruction formats in improving recommendation quality and user interaction.

4.3 Fine-Tuning LLMs for Recommendation

Instruction fine-tuning for recommendation can be implemented as follows:

1. **Dataset Curation**: Construct a collection of diverse instruction-response pairs. Each pair simulates a user intent followed by an appropriate system recommendation.

Example:

```
Instruction: "Suggest a budget-friendly Italian restaurant
in downtown."
Response: "Pasta Palace, Bella Italia."
```

2. **Fine-Tuning**: Train the LLM on these pairs using supervised learning. The model learns to map natural language instructions to appropriate recommendations based on context and content.
3. **Deployment**: Once fine-tuned, the model can be used in production to interpret incoming user queries, making recommendations on the fly without needing explicit rules or templates.

4.3.1.1 Task Types

When designing instruction-based fine-tuning for recommendation models, it is crucial to account for both the *user context* and the *form of the task* being modeled. User context can range from vague exploratory intent to highly specific requests and it influences how preferences are expressed. Task form determines the structure of the model's output, such as evaluating, comparing, or ranking items.

1. **User Context**:
   - **Cold-Start/Exploratory**: General queries (e.g., "Recommend a popular sci-fi movie").
   - **Contextual/Vague**: Implicit or partially defined preferences (e.g., "Suggest a nearby coffee shop").
   - **Explicit/Specific**: Clear, detailed instructions (e.g., "Find a romantic comedy with a happy ending").
2. **Task Form**:
   - **Pointwise**: Evaluate single items (e.g., "Is this movie kid-friendly?").
   - **Pairwise**: Compare items (e.g., "Which is better: Inception or Interstellar?").
   - **Matching and Reranking**: Retrieve and rank items (e.g., "Rank these restaurants by ambiance").

4.3.1.2 Benefits

- **Flexibility**: Handles diverse tasks without task-specific architectures.
- **User Control**: Enables natural language queries for personalized recommendations.
- **Reduced Prompt Engineering**: Learns to interpret instructions inherently.

4.3.2 Domain Knowledge Fine-Tuning

While general-purpose LLMs exhibit strong zero-shot performance, they often struggle to deliver high-quality recommendations in specialized domains such as fashion retail, automotive shopping, or financial services. To address this, *domain-adaptive pre-training (DAPT)* has emerged as a promising strategy, where a pre-trained language model is further trained on unlabeled domain-specific corpora to better align with specialized vocabulary, semantics, and discourse patterns (Gururangan et al., 2020). Another approach, *domain knowledge fine-tuning*, offers more targeted adaptation by adjusting the model's parameters using supervised signals specific to the target domain.

This section focuses on *domain knowledge fine-tuning*, which involves adapting a foundational model to the language, structure, and behavioral cues of a specific vertical by training it on curated, domain-relevant data. The objective is to align the model's internal representations with the unique semantics and user interaction patterns of that domain, thereby enhancing the model's recommendation accuracy, interpretability, and personalization.

4.3.2.1 Implementation and Examples

Domain-specific fine-tuning follows the same supervised fine-tuning (SFT) process outlined in Chap. 1, which includes collecting data, formatting prompt–response pairs, and updating model weights. What distinguishes it is the construction of training data, which must capture domain-specific language, context, and user interaction patterns.

- **Fashion**: Data from style guides, product descriptions, and fashion blogs enables models to generate personalized advice like "layered minimalist outfits for cold weather" or "gender-neutral capsule wardrobe essentials."
- **Automotive**: Reviews, specs, and forum discussions allow models to interpret nuanced queries such as "a compact car for urban driving" vs. "a hybrid SUV for long trips," offering targeted suggestions with feature trade-offs.
- **Finance**: Articles labeled by topic or sentiment, plus engagement data, help models recommend content aligned with user goals, for example, "tax-efficient retirement planning" or "emerging market ETF risks."

4.3 Fine-Tuning LLMs for Recommendation   111

The goal is not data volume, but embedding domain-relevant signals for better reasoning and grounded recommendations.

4.3.2.2 Benefits

Domain fine-tuning offers key advantages for recommendation systems:

- **Semantic Precision**: Improves understanding of niche terms (e.g., "boho-chic" or "monochrome layering" in fashion).
- **Better Personalization**: Adapts to user intent, such as tailoring finance content for beginners vs. experts.
- **Cold-Start Mitigation**: Leverages domain semantics to serve relevant suggestions even with minimal user history.

These benefits help address common challenges like sparse data, weak personalization, and semantic mismatch.

4.3.3 Personalized LLM Fine-Tuning

Personal preference fine-tuning refers to the process of adapting a language model to the tastes and behavioral patterns of an individual user or a small user segment. Unlike domain-level fine-tuning, which generalizes across a category (e.g., fashion or finance), personal preference fine-tuning aims to capture hyper-personalized signals from a specific user's interaction history, preferences, and goals. The objective is to enable recommendation systems that respond not just to general patterns, but to each user's unique style, intent, and context.

4.3.3.1 Implementation

The fine-tuning process begins by collecting user-specific interaction data. This can include:

- Browsing history and session logs.
- Past purchases, likes, and ratings.
- Explicit inputs such as favorite genres, budget preferences, or aspirational goals.

This data is then formatted into *structured prompt–response pairs* or sequences to fine-tune the LLM in a supervised manner. For instance, prompts like:

```
Prompt: "The user recently liked [Product A], [Product B],
and [Product C]. What should be recommended next?"
Response: "[Product D], because it shares features with A
and is popular among similar users."
```

This process updates the LLM's internal representation space to *encode user behavior patterns*, going beyond template prompting by instilling preferences directly into the model's parameters.

In practice, many systems adopt *user-segment-level fine-tuning* to balance personalization and scalability. Instead of fine-tuning per user, models are adapted for segments like:

- "Budget-conscious shoppers"
- "Tech-savvy early adopters"
- "Frequent travelers"

This enables more generalizable and reusable personalization while retaining behavior-aware benefits.

4.3.3.2 Choice Between Personalized User Embedding and Personalized LLM Fine-Tuning

There are two dominant paradigms for user adaptation: personalized embedding injection and personalized LLM fine-tuning. Table 4.2 compares between personalized embedding and LLM fine-tuning in terms of scalability, update frequency, personalization depth, etc.

Personalized Embedding Injection

- Keeps the base LLM frozen
- Injects user or item embeddings at runtime (via prefix tuning, adapters, or contextual embedding layers)
- Embeddings can be learned via lightweight methods (e.g., matrix factorization, CLIP-style projection)

Table 4.2 Key decision factors between personalized embeddings and LLM fine-tuning

Criteria	Personalized embeddings	LLM Fine-Tuning
Scalability	Efficient at large scale	Costly and complex
Update frequency	Embeddings can refresh in real time	Retraining is slow
Personalization depth	Shallow (structural info only)	Deep (stylistic and semantic match)
Privacy and compliance	Easier anonymization	Risk of data retention in weights
Ideal use case	Broad consumer platforms (e.g., e-commerce, news)	High-value users or specialty domains

4.3 Fine-Tuning LLMs for Recommendation

Pros

- Lightweight and fast updates
- Anonymization-friendly (GDPR-safe)
- Ideal for dynamic environments (e.g., fast fashion, real-time recommendations)

Cons

- Limited personalization depth and may miss subtle user preferences
- Requires thoughtful integration into model architecture

Personalized LLM Fine-Tuning

- Retrains or adapts the full model (or LoRA modules) using user or segment-specific data

Pros

- Deep personalization—captures user tone, interests, and context nuances
- Superior for high-stakes or niche domains (e.g., financial advising, luxury shopping)
- Enhances recommendation fluency and rationale generation

Cons

- Computationally expensive (especially per user)
- Privacy concerns—user data is encoded into model weights
- Challenging to update frequently as preferences evolve

We summarize the key decision factors between personalized embeddings and personalized LLM fine-tuning in Table 4.2. As a practical guidance:

- Use *embedding-based personalization* for real-time, privacy-safe, and scalable recommendations.
- Use *fine-tuning* for high-value users, long-lived preferences, or domains where recommendation quality outweighs cost.
- A hybrid approach is often most effective: Use embeddings as default and fine-tune for VIP segments or high-value contexts.

4.3.4 Summary and Discussion

To summarize, we present the objectives, data used, scope, and granularity of each fine-tuning technique in Table 4.3.

- *Fine-tuning for domain knowledge* is widely used, as we can fine-tune LLMs to specialized domains like healthcare, finance, etc. Lots of commercial applications.

Table 4.3 Fine-tuning techniques for recommendation

Aspect	Fine-tune for recommendation instruction	Fine-tune for domain knowledge	Fine-tune for personal preferences
Scope	Adapt the model to follow structured instructions across diverse users and contexts	Incorporate specialized knowledge for accurate recommendations in a specific domain (e.g., healthcare, academia)	Deliver highly personalized recommendations tailored to individual users' behavior and preferences
Data used	Task-specific labeled data with recommendation instructions (e.g., user-item pairs, ranking tasks)	Domain-specific corpora such as product catalogs, scientific papers, or industry datasets	Individual user data, including interaction history, ratings, or personal notes, often with privacy safeguards
Granularity	Task-level adaptation applicable across multiple domains	Domain-specific adaptation focused on industry or topic area	Highly granular personalization based on individual user signals

- *Fine-tuning for recommendation tasks* is specific to recommendations, we may also need to augment LLM with domain knowledge (like movies, e-commerce products) for it to better perform certain tasks.
- *Fine-tuning for personal preferences* is cutting edge, as part of the efforts of creating AI persona.

While fine-tuning is effective, several alternatives address its limitations or provide complementary benefits:

1. **Prompt Engineering**: Pre-trained LLMs like GPT-4 can perform recommendation tasks without additional fine-tuning by leveraging well-crafted prompts. By including *few-shot examples*, *structured user preferences*, and even *reasoning paths* (e.g., "because the user liked X, they may enjoy Y"), we can guide the model to exhibit desired behaviors. This approach enables flexible and zero-shot or few-shot personalization while avoiding the overhead of task-specific retraining.
2. **Retrieval-Augmented Generation (RAG)**: RAG combines LLMs with external knowledge retrieval, enabling models to access up-to-date or domain-specific information without fine-tuning (Borgeaud et al., 2022). For instance, a recommendation system can retrieve relevant product details from a database to enhance response accuracy.
3. **Reinforcement Learning with Human Feedback (RLHF)**: RLHF refines models based on iterative human evaluations, addressing SFT's limitations in incorporating preferential feedback and aligning outputs with user expectations.

4.4 Production-Ready Optimization for LLM as Recommender

In this chapter, we will explore how to design workflows that capitalize on these capabilities, focusing on how LLMs can serve as the foundation for next-generation recommendation systems.

Deploying LLM-based recommendation systems at scale often involves significant computational and financial costs. To optimize these costs while maintaining system performance, techniques such as model distillation and quantization are employed. These methods have been effectively utilized in industry applications to enhance efficiency in training and deployment of LLMs (Behdin et al., 2025).

This section explores three key techniques—*model distillation, quantization,* and *caching*—to optimize costs while maintaining system performance, as illustrated in Fig. 4.2.

For optimal cost-efficiency, *distillation, quantization*, and *caching* can be strategically combined:

- *Use distilled and quantized models* for real-time, personalized queries where flexibility and low latency are essential. Quantization reduces memory footprint and inference time, enabling deployment on resource-constrained infrastructure without sacrificing much accuracy.
- *Leverage caching* for frequently accessed or static data, such as trending items or popular categories, to avoid redundant computation and minimize latency.

For example, a movie recommendation system might use a distilled model to generate personalized suggestions for individual users while caching results for trending movies or frequently searched genres.

4.4.1 Knowledge Distillation

Knowledge distillation is a model compression technique in which a smaller, lightweight model (the *student*) is trained to replicate the behavior of a larger, high-performing model (the *teacher*) (Hinton et al., 2015). Originally introduced to improve the deployment efficiency of deep learning systems, distillation has become

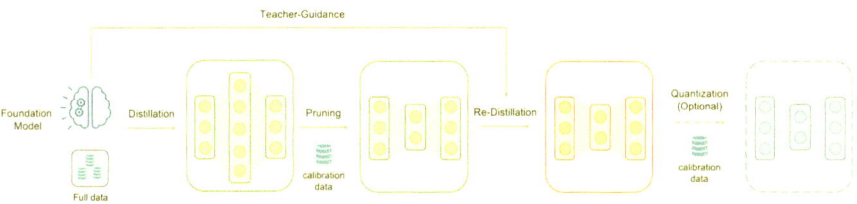

Fig. 4.2 Overview of the process of creating deployable models via distillation and compression

a practical solution for building scalable recommendation systems—especially when using LLMs like GPT-4 or FLAN-T5-Large is too resource-intensive for real-time inference.

In recommendation systems, knowledge distillation enables developers to preserve the strengths of powerful LLMs while reducing computational cost, inference latency, and memory footprint.

4.4.1.1 Knowledge Distillation Implementation

Knowledge distillation compresses a large, high-performing *teacher model* into a smaller, efficient *student model*, preserving key behaviors while improving deployability. The typical pipeline consists of three main steps:

1. **Choosing Teacher–Student Pair**. Select a teacher–student model pair suited to your task and deployment constraints. For example, a teacher could be GPT-4 or FLAN-T5-Large, while the student might be DistilBERT or FLAN-T5-Small.
2. **Teacher Inference to Generate Soft Targets**. The teacher processes input data (e.g., user histories, item features) and outputs logits—unnormalized scores representing its internal beliefs across all output classes.
3. **Student Model Training**. The student is trained to match the teacher's softened output distributions, rather than only learning from ground truth labels. The training loss typically minimizes divergence (e.g., KL divergence) between the student and teacher distributions.

4.4.1.2 Benefits

- **Efficiency**: Significant reduction in memory footprint and computational cost, and easier to deploy on edge devices or low-resource environments.
- **Low Latency**: Student models enable real-time recommendation in production systems.
- **Retained Performance**: With well-executed distillation, student models often maintain performance within 5–10% of the teacher model on many tasks.

4.4.1.3 Challenges

- **Training Overhead**: Initial distillation requires computational resources and access to teacher inference outputs.
- **Loss Function Configuration**: Distillation can be challenging to tune—student models may miss the teacher's nuanced reasoning without carefully designed loss functions.

4.4.1.4 Best Practices and Considerations

- **Using Logits and Soft Targets**:
  - **Preservation of Relative Class Information**: Unlike hard one-hot labels, logits retain information about how the teacher ranks all options. For instance, logits [2.0, 1.0, 0.2] (before softmax) implies strong relative preference that help the student model learn the distinction between items.
  - **Smoother Learning via Soft Targets**: Soft targets, generated via temperature scaling, encourage smoother gradients during training. This leads to better generalization and avoids overfitting to hard decisions.
  - **Distribution Alignment**: Distilling via logits aligns the full output distribution between teacher and student, facilitating a deeper behavioral match than using hard labels alone.
  - **Logit Standardization**: More recent techniques, such as logit standardization, normalize differences in scale between teacher and student logits, improving training stability and performance transfer.
- **Combining Soft and Hard Targets**. In practice, it's common to use a *blended loss*—a weighted combination of: distillation loss (soft targets) and supervised loss (hard labels). This hybrid approach encourages both accurate predictions and robust generalization.
- **Post-Distillation Fine-Tuning**. After initial distillation, student models are often fine-tuned on downstream data—like user-item interactions or session sequences—to adapt to domain-specific recommendation needs.

4.4.2 Quantization and Model Compression

This section explores *quantization* and *model compression*, two critical techniques for optimizing the cost and efficiency of large language model (LLM)-based recommendation systems. We discuss their definitions, motivations, and practical implementations, providing a concise guide for researchers and practitioners.

Quantization reduces the precision of model weights and activations, typically from 32-bit floating-point (FP32) to lower precision formats like 16-bit floating-point (FP16) or 8-bit integers (INT8). Model pruning, another compression technique, removes redundant or less important weights to reduce model size and inference cost. Recent advances like QLoRA (Dettmers et al., 2022) demonstrate how quantization can be combined with low-rank adaptation to enable memory-efficient fine-tuning of large models at scale.

Quantization and model compression are essential for reducing computational costs, improving scalability, and enhancing energy efficiency. They enable deployment on resource-constrained devices, such as mobile phones and edge devices, and complement other cost optimization methods like distillation, caching, and response reuse.

4.4.2.1 Techniques and Software Packages

Techniques

1. **Quantization**:
   - **Post-Training Quantization**: Applied after training to quantize weights and activations.
   - **Quantization-Aware Training**: Optimizes the model during training to account for quantization.
   - **Dynamic Quantization**: Quantizes weights and activations dynamically during inference.

2. **Model Compression**:
   - **Pruning**: Removes less important weights or neurons.
   - **Knowledge Distillation**: Trains a smaller student model to mimic a larger teacher model.
   - **Low-Rank Factorization**: Approximates weight matrices with lower rank representations.

Software Packages

- **TensorFlow Lite**: Provides tools for post-training quantization and quantization-aware training, enabling efficient deployment on mobile and edge devices.
- **PyTorch**: Supports dynamic quantization, quantization-aware training, and pruning, making it versatile for model optimization.
- **Hugging Face Transformers**: Offers pre-trained models and tools for knowledge distillation, simplifying the creation of smaller, efficient models.
- **ONNX Runtime**: Optimizes models for inference with quantization and pruning, ensuring high performance across platforms.

4.4.3 Caching and Response Reuse

While knowledge distillation addresses the cost and latency of deploying large models by compressing them into smaller variants, another powerful technique is *caching*—the reuse of previously computed outputs to avoid redundant inference. Caching strategies are particularly effective in large-scale LLM-based recommendation systems, where many user queries or content scenarios recur across sessions, users, or platforms.

Caching involves storing outputs or intermediate representations generated by the LLM so that future requests can be served directly from memory or disk, without invoking the full inference pipeline. This technique is especially useful in reducing computation cost, improving system responsiveness, and scaling real-time applications under high user load. Caching is especially effective in scenarios where:

- *User behavior exhibits repetition*, such as revisiting categories, sessions, or saved searches.
- *High-traffic content* (e.g., popular books, movies, or fashion items) is recommended repeatedly to different users.
- *Inference cost is high*, and freshness of recommendations is less critical than response speed or cost.

By reducing the number of direct model calls, caching enables systems to *scale affordably*, even when using large LLMs behind the scenes.

4.4.3.1 Benefits

- **Significant cost savings**: Reduces reliance on expensive GPU-based inference or paid API usage.
- **Improved latency**: Serving cached responses is substantially faster than live inference.
- **Enhanced scalability**: Allows the system to support more users with fewer resources.

4.4.3.2 Limitations

- **Staleness of Results**: Cached responses can become outdated if user interests change or item catalogs are frequently updated.
- **Cache Management Complexity**: Effective caching requires thoughtful strategies for cache invalidation, expiration, and refresh.
- **Storage Overhead**: Storing large volumes of embeddings or prompt–response pairs can consume memory or disk space, requiring optimization.

4.4.3.3 Caching Strategy and Best Practice

Table 4.4 summarizes the common types of caching in the context of recommendation systems:

- **Prompt–Response Outputs**: Reusing previously generated recommendations for common queries or conversational patterns.
- **User and Item Embeddings**: Caching vector representations that are used in candidate retrieval or similarity search.
- **Intermediate Pipeline Outputs**: Retaining tokenized inputs, attention maps, or reranking scores that are costly to recompute.

Here are some best practices for designing caching:

- **Implement cache keys based on query fingerprinting or user-context hashes**. To achieve this, we can generate unique cache keys by hashing a normalized ver-

Table 4.4 Different caching types (prompt-response caching, embedding caching and intermediate output caching) and their typical use cases

Caching type	Description	Typical use cases
Prompt–response caching	Stores final LLM-generated outputs (e.g., product suggestions or article lists)	High-frequency queries; FAQs; chatbot-style recommenders
Embedding caching	Stores dense vectors for users/items for fast retrieval or reranking	Vector search systems; hybrid recommenders
Intermediate output caching	Stores tokenized inputs or partial model computations	Multi-stage recommendation pipelines; reranking modules

sion of the input query or combining it with relevant user context (e.g., location, preferences). This ensures that semantically similar inputs retrieve consistent results and avoids redundant computation for frequently asked or behaviorally similar queries.

- **Define refresh policies for dynamic content domains (e.g., daily or hourly regeneration)**: For content that changes regularly—like news, stock data, or trending items—establish cache expiration rules. For instance, regenerate recommendation candidates every hour for fast-moving domains like social media, or once per day for e-commerce platforms with slower changing catalogs.
- **Use embedding versioning to track compatibility across model updates**: Introduce a version control system for embedding models, appending version tags to cache keys or metadata. This ensures that stale embeddings from older models don't pollute the results when the underlying model is updated, maintaining consistency and avoiding compatibility issues.
- **Combine caching with knowledge distillation, using a lightweight model to handle uncached queries efficiently**: Use a distilled or compressed version of the main LLM to serve cache misses. This speeds up inference for first-time queries while reducing infrastructure costs. The full LLM can still handle complex or critical queries selectively, preserving quality without sacrificing scalability.

4.4.4 Design Trade-Offs and Practical Considerations

Designing workflows for LLM-based recommendation systems involves balancing several critical factors to ensure efficiency, scalability, and high-quality recommendations. Below are the five most salient considerations, along with actionable recommendations:

4.4.4.1 Cost Vs. Quality Vs. Latency

- **Consideration**:
  - **Cost**: Larger LLMs (e.g., GPT-4) are computationally expensive, while smaller models are more cost-effective.
  - **Quality**: Larger models generally produce higher quality recommendations due to their superior contextual understanding and reasoning capabilities.
  - **Latency**: Larger models often have higher inference times, which can be problematic for real-time applications. Smaller models are faster but may sacrifice quality.

- **Recommendation**:
  - Use smaller models or caching for routine tasks (e.g., candidate generation) to reduce costs and latency.
  - Reserve larger models for complex queries requiring high-quality, nuanced recommendations (e.g., personalized suggestions).
  - Optimize for latency by pre-filtering candidates or using embeddings for efficient retrieval, ensuring real-time responsiveness without compromising quality.

4.4.4.2 Fine-Tuning Vs. Retrieval-Augmented Generation (RAG)

- **Consideration**:
  - **Fine-Tuning**: Tailors LLMs to specific domains but requires extensive labeled data and computational resources.
  - **RAG**: Dynamically incorporates external knowledge, offering flexibility but adding system complexity.

- **Recommendation**:
  - Use fine-tuning for stable domains with well-defined tasks (e.g., movie recommendations).
  - Use RAG for dynamic or rapidly evolving use cases (e.g., news or trending product recommendations).

4.4.4.3 Self-Built Models Vs. APIs

- **Consideration**:
  - **Self-Built Models**: Provide greater control and long-term cost-efficiency but require significant infrastructure and expertise.
  - **APIs**: Enable rapid deployment and scalability but increase ongoing costs and dependency on external providers.

- **Recommendation**:
  - Start with APIs for prototyping and small-scale deployments.
  - Transition to self-built models for production-scale systems to reduce costs and improve control.

4.4.4.4 Prompt-Driven Versatility and Minimizing Dependencies

- **Consideration**: Well-designed prompts can replace multiple traditional components (e.g., retrieval, ranking, explanation generation), simplifying the architecture and reducing dependencies.
- **Recommendation**:
  - Leverage instruction-based prompting to consolidate tasks into a single LLM query.
  - Avoid unnecessary systems (e.g., vector search engines) unless critical to performance.
  - Continuously refine prompts to improve recommendation quality and reduce the need for additional modules.

4.4.4.5 Scalability and Cost-Efficiency

- **Consideration**: For large-scale systems, additional mechanisms (e.g., pre-filtering, embeddings) are needed to ensure efficiency and manage costs.
- **Recommendation**:
  - Use embeddings and approximate nearest neighbor (ANN) search for efficient candidate retrieval.
  - Cache frequent queries and summarize inputs to minimize token length and API costs.
  - Combine LLMs with traditional recommendation techniques (e.g., collaborative filtering) to handle scalability challenges.

4.5 Tutorial: Fine-Tuning LLMs for Personalized Movie Recommendations

4.5.1 Overview

This tutorial explores fine-tuning Large Language Models (LLMs) for personalized movie recommendation using the MovieLens dataset. The objective is to predict future movie preferences based on a user's prior ratings and movie metadata (e.g., genre). We compare fine-tuned LLMs with few-shot prompting to assess their effectiveness, efficiency, and scalability.

Key Concepts

- **Fine-Tuning**: Adapting pre-trained LLMs to recommendation tasks using user-item interaction data.
- **Few-Shot Prompting**: Leveraging LLMs with minimal task-specific examples to generate recommendations.
- **Evaluation**: Measuring recommendation quality using precision, recall, and computational efficiency.

Goal of the Tutorial

- Learn how to structure a recommender dataset as a natural language sequence suitable for text-to-text models like Flan-T5.
- Understand the process of distilling a large language model into a smaller one using logit regression and MSE loss.
- Gain hands-on experience comparing teacher and student model outputs to evaluate the effectiveness of distillation.

We show a condensed version of this tutorial in the book text. The full version of the code is available at: https://github.com/qqwjq1981/springer-LLM-recommendation-system

4.5.2 Experimental Design

4.5.2.1 Dataset Preparation

- **Data Source**: MovieLens 1M dataset.
- **Labeling**: Movies rated 4–5 stars are labeled as "liked," 1–2 stars as "disliked."
- **Target Generation**: Only high-rated future movies are used for evaluation.
- **Metadata**: Genre and release year are added to movie descriptions.

4.5.2.2 Prompt Construction

- **Input Format**:

```
User liked: [list of liked movies with genres].
User disliked: [list of disliked movies with genres].
Recommend new movies. Do not repeat any from history.
Output movie titles separated by semicolons.
```

- **Target**:

```
{"recommended_movies": ["Movie A", "Movie B", ...]}
```

4.5.2.3 Fine-Tuning Approach

- **Model**: We evaluate sub-1B parameter models and select amd/AMD-OLMo-1B-SFT for its strong out-of-box performance.
- **Training**: Fine-tune the model to predict future liked movies.
- **Parameter-Efficient Fine-Tuning (PEFT)**: Use LoRA to reduce compute and memory usage.
- **Baselines**:
  - *Zero-Shot:* Use the model without adaptation.
  - *Few-Shot:* Use 5 in-context examples selected at random.

4.5.2.4 Evaluation

- **Metrics**: Precision@K, Recall@K, NDCG@K (K = 5, 10, 20).
- **Test Set**: Held-out interactions per user.

4.5.3 Results and Analysis

Table 4.5 illustrates evaluation metrics to compare the performance of zero-shot, few-shot and fine-tuning for movie recommendation:

1. **Fine-tuning yields the strongest performance**. Fine-tuned models significantly outperform zero-shot and few-shot baselines across all metrics. Precision@5 more than doubles compared to zero-shot (0.1553 vs. 0.0633), and the best NDCG@10 score (0.3133) indicates improved ranking quality.

Table 4.5 Evaluation metrics (precision, recall, and NDCG) for zero-shot, few-shot, and fine-tuning for movie recommendation

Approach	Metric	@5	@10	@20
Zero-shot	Precision	0.0633	0.0583	0.0365
	Recall	0.0045	0.0074	0.0091
	NDCG	0.1073	0.0673	0.0087
Few-shot	Precision	0.0940	0.0913	0.0953
	Recall	0.0071	0.0142	0.0337
	NDCG	0.2426	0.2867	0.3260
Fine-tuning	Precision	0.1553	0.1593	0.1422
	Recall	0.0114	0.0227	0.0410
	NDCG	0.2981	0.3133	0.2210

2. **Few-shot prompting is a strong, efficient baseline**. Despite no model updates, few-shot prompting delivers substantial gains over zero-shot (e.g., NDCG@20: 0.3260 vs. 0.0087), making it a practical choice for cold-start or low-resource settings.
3. **Recall remains low due to generative flexibility**. All methods show low recall, which is expected given the open-ended nature of generation. Models often produce reasonable recommendations not in the ground truth. Hence, recall should be viewed as a lower bound and ideally supplemented with human or implicit feedback.

4.5.4 Conclusion

4.5.4.1 Recommendations

- Use *LoRA fine-tuning* for production systems requiring accuracy and personalization.
- Adopt *few-shot prompting* for rapid prototyping or when compute resources are limited.
- In *cold-start scenarios*, begin with few-shot learning and transition to fine-tuning as user data accumulates.

4.5.4.2 Key Takeaways

- Fine-tuned LLMs outperform prompting-based approaches in accuracy and relevance.
- Few-shot prompting offers a scalable, training-free alternative with acceptable diversity.
- Metadata such as genre and release year substantially improves recommendation quality.

4.6 Second Tutorial: Knowledge Distillation Using MovieLens Dataset

4.6.1 Overview

In this tutorial, we demonstrate knowledge distillation using the MovieLens 1M dataset, where we distill the Flan-T5-Large model (teacher) into the Flan-T5-Small model (student). The goal is to predict movie ratings based on movie titles and genres, leveraging sequential data from the MovieLens dataset.

The key steps in this tutorial include:

1. **Dataset Preparation**: Constructing a sequential dataset from MovieLens 1M for a text-based recommendation task.
2. **Model Distillation**: Distilling the Flan-T5-Large model into the Flan-T5-Small model using logits and Mean Squared Error (MSE) loss.
3. **Evaluation**: Comparing the performance of the distilled student model with the teacher model.

Goal of this Tutorial

- Organize a recommendation dataset into a format suitable for text-to-text models.
- Perform knowledge distillation using logit-based regression with MSE loss.
- Evaluate and compare the performance of teacher and student models.

We show a condensed version of this tutorial in the book text. The full version of the code is available at: https://github.com/qqwjq1981/springer-LLM-recommendation-system

4.6.2 Experimental Design

4.6.2.1 Dataset Preparation

The MovieLens 1M dataset consists of 1 million movie ratings from users, along with movie metadata (titles and genres). We preprocess the data to create a sequential dataset where the task is to predict the rating based on the movie title and genres.

We use the gpt-4o-mini model to filter out hard examples in the tutorial by automatically identifying and removing samples that the model answers incorrectly. This ensures that the distilled model is trained primarily on examples that are reliably understood by a strong teacher model, improving label quality and training stability.

The filtered dataset is then split into training and testing sets (80/20 split), and we consider two prompt versions:

- Without chain-of-thought reasoning

```
prompt:  The user liked the following movies: Fargo (1996),
Antz (1998), Airplane! (1980).
    Which movie is the user more likely to prefer?
    1. Thomas Crown Affair, The (1968) (Crime|Drama|Thriller)
    2. Bambi (1942) (Animation|Children's)
Please answer with 1 or 2 only.
label:   2
decoded:   2
```

4.6 Second Tutorial: Knowledge Distillation Using MovieLens Dataset

- With zero-shot chain-of-thought reasoning:

```
prompt:  The user liked the following movies: Fargo (1996),
Antz (1998), Airplane! (1980).
Please think step-by-step about the genre and the year of
each movie when making a decision.
    Which movie is the user more likely to prefer?
    1. Thomas Crown Affair, The (1968) (Crime|Drama|Thriller)
    2. Bambi (1942) (Animation|Children's)
Please answer with 1 or 2 only.
label:  2
decoded:  2
```

4.6.2.2 Teacher and Student Models

- **Teacher Model**: Flan-T5-Large (783M parameters).
- **Student Model**: Flan-T5-Small (77M parameters).

4.6.2.3 Distillation Process

1. **Training Setup**: The teacher model generates logits (raw predictions) for the training data. The student model is trained on the MovieLens dataset using the teacher's logits as soft targets.
2. **Loss Function**: We considered two alternative loss functions:
   (a) KL divergence between the logits of the teacher and student models (KL).
   (b) Hybrid distillation loss that combines KL divergence with cross-entropy loss (KL + Cross-Entropy), similar to Behdin et al. (2025).

4.6.2.4 Evaluation Metrics

- **Accuracy**: The primary metric for evaluating the performance of both the teacher and student models.
- **Efficiency**: Inference time and model size are compared to highlight the trade-off between performance and efficiency.

4.6.3 Results and Analysis

Table 4.6 summarizes the performance of the teacher and student models and time cost on training and inference. We skip the results using CoT prompting and only present those without CoT prompting:

Table 4.6 Comparison between teacher model, student model, and distilled model in accuracy and inference time

Variation	Teacher Acc (flan-T5-large)	Student Acc (flan-T5-small)	Distilled Acc	Teacher inference time per sample (ms)	Student inference time per sample (ms)	Distillation time on CPU (h)
No CoT, KL	60.3%	56.1%	55.7%	1251	130	2.7
No CoT, Hybrid	60.3%	56.1%	53.8%	1231	128	3.2

1. **Efficiency Gains**
   Despite the limited accuracy improvements, distillation still offers substantial efficiency benefits. The student model is over *10× faster* (≈130 ms vs. 1250 ms per sample) and *10× smaller* (77M vs. 783M parameters), making it highly suitable for deployment in resource-constrained environments. Although distillation incurs a one-time cost (~3 CPU hours across runs), the long-term inference efficiency gains are significant.
2. **Distillation Performance and Limitations**
   In our current setup, the distilled student model does *not consistently outperform* the base Flan-T5-Small; in some cases, accuracy slightly drops (e.g., 55.7% vs. 56.1% in No CoT, KL setting). This suggests that the effectiveness of distillation may be constrained by data quality, model capacity, or the lack of a diverse and challenging enough training signal. These results highlight the need for more refined distillation strategies (e.g., better filtering or stronger supervision) to realize meaningful gains.
3. **Hybrid Loss Function Insights**
   Contrary to expectations, the hybrid loss function *did not outperform* standard KL divergence in this evaluation. In fact, distilled accuracy slightly dropped with hybrid loss (53.8% vs. 55.7%), possibly due to over-regularization or interference between objectives. While hybrid loss remains a promising direction, these early results suggest it requires further tuning and larger scale validation to assess its full potential.

4.7 Conclusions

This tutorial illustrates the application of *large language model distillation* in the context of recommendation systems, using pairwise preference prediction on the MovieLens 1M dataset. By distilling Flan-T5-Large into a lightweight Flan-T5-Small student, we demonstrate how to build *faster and smaller models* that retain much of the teacher's reasoning capabilities.

Despite the numerical results of distillation still has room to improve, the tutorial provides critical insights into the *distillation pipeline* and *loss function design*.

These results reflect the *inherent difficulty* of pairwise recommendation tasks when using only implicit feedback and limited supervision.

This work sets the stage for more advanced follow-ups, such as:

- *Hyperparameter tuning* (e.g., distillation temperature, margin losses)
- *Task-specific fine-tuning* on richer user-item datasets
- *Evaluation with ranking metrics* to go beyond binary accuracy
- *Deployment validation* to assess real-world recommendation effectiveness

References

Behdin, K., Dai, Y., Fatahibaarzi, A., et al. (2025). *Efficient AI in practice: Training and deployment of efficient LLMs for industry applications.* arXiv:2502.14305.

Borgeaud, S., Mensch, A., Hoffmann, J., et al. (2022). *Improving language models by retrieving from trillions of tokens.* arXiv:2112.04426.

Dettmers, T., Pagnoni, A., Holtzman, A., & Zettlemoyer, L. (2022). *QLoRA: Efficient fine-tuning of quantized LLMs.* arXiv:2305.14314.

Geng, Z., Wu, L., Liu, C., et al. (2023). *OpenP5: Benchmarking foundation models for recommendation.* arXiv:2306.11134.

Gururangan, S., Marasović, A., Swayamdipta, S., et al. (2020). *Don't stop pretraining: Adapt language models to domains and tasks.* ACL.

Hinton, G., Vinyals, O., & Dean, J. (2015). *Distilling the knowledge in a neural network.* arXiv:1503.02531.

Hu, E. J., Shen, Y., Wallis, P., et al. (2021). *LoRA: Low-rank adaptation of large language models.* arXiv:2106.09685.

Wu, L., Zheng, Z., Qiu, Z., Wang, H., Gu, H., Shen, T., Qin, C., Zhu, C., Zhu, H., Liu, Q., Xiong, H., & Chen, E. (2023). *A survey on large language models for recommendation* (v5). arXiv.

Xu, L., Zhang, J., Li, B., Wang, J., Cai, M., Zhao, W. X., & Wen, J.-R. (2024). *Prompting large language models for recommender systems: A comprehensive framework and empirical analysis.* arXiv preprint arXiv:2401.04997.

Zhang, J., Xie, R., Hou, Y., Zhao, W. X., Lin, L., & Wen, J.-R. (2023). *Recommendation as instruction following: A large language model empowered recommendation approach.* arXiv.

Chapter 5
Conversational Recommendation Systems

This chapter introduces conversational recommendation systems (CRS), focusing on the integration of reinforcement learning (RL) and large language models (LLMs) to enable dynamic, interactive recommendations. It starts by outlining foundational RL algorithms such as multi-armed bandits, deep Q-networks, and policy gradients. The chapter then discusses RL and LLM applications in dialogue management, personalization, and reward design. The chapter also details key CRS modules, including intent detection, state tracking, clarification mechanisms, and evaluation strategies. A practical tutorial demonstrates how to build a CRS using RL and LLMs, offering insights into system design, preference extraction, and reward modeling for real-world deployment.

5.1 Reinforcement Learning Foundations for Conversational Recommendation

5.1.1 Introduction

Reinforcement Learning (RL) is a machine learning paradigm in which an agent learns to make decisions by interacting with an environment and receiving feedback in the form of rewards (Sutton and Barto 2018). The core objective is to learn a policy denoted as ($\pi(a|s)$) that maps each state (s) to an action (a) in order to maximize the cumulative reward over time. Unlike *supervised learning*, where models learn from labeled data, or *unsupervised learning*, which uncovers patterns in unlabeled data, RL learns through trial-and-error, guided by delayed and often sparse rewards.

Key components of RL include:

- **State (s)**: The representation of the current context, such as a user's profile, preferences, or dialogue history.
- **Action (a)**: A decision the agent makes, such as recommending an item, asking a clarifying question, or suggesting a product bundle.
- **Reward (r)**: A feedback signal indicating how good or bad an action was, often inferred from user behavior (e.g., clicks, purchases, dwell time).
- **Policy (π)**: The decision-making strategy of the agent.
- **Value function (V or Q)**: Estimations of expected future rewards for each state or state-action pair.

Traditional recommendation algorithms often rely on static user-item interaction histories and do not adapt well to dynamic contexts. In contrast, RL-based recommenders actively learn from sequential user interactions and optimize for long-term outcomes. RL provides a framework for optimizing long-term user satisfaction, engagement, or conversion rather than immediate rewards. It enables systems to continuously learn and adapt based on user feedback, making it particularly well-suited for interactive and conversational recommendation scenarios.

For example, in a multi-turn conversational setting, an RL-based agent can learn when to recommend an item, ask for more user preferences, or switch domains altogether. In this spirit, an e-commerce chatbot can learn to strategically alternate between offering promotions and gathering user preferences.

5.1.2 Types of RL Algorithms in Recommendation

In this section, we introduce several core RL algorithms: Multi-Armed Bandits, Deep Q-Network, Policy Gradient, and Monte Carlo Tree Search. For each algorithm, we discuss how the algorithm works, common techniques and its use case in recommendation systems.

5.1.2.1 Multi-Armed Bandit (MAB)

MAB algorithms are a class of RL methods that address the exploration-exploitation trade-off in environments with a fixed set of actions and immediate feedback (Lattimore & Szepesvári 2020). They are simple, efficient, and widely used in real-time recommendation scenarios.

How it works: The system repeatedly selects from a fixed pool of items (arms) and receives immediate reward signals (e.g., click, purchase). The objective is to maximize cumulative reward over time by balancing exploration (trying new items) and exploitation (recommending known high-reward items).

Common algorithms:

- **ε-greedy**: Chooses the best-known item with probability $1 - \varepsilon$ and explores randomly with probability ε.
- **Upper Confidence Bound (UCB)**: Selects items based on the sum of estimated reward and a confidence interval, encouraging exploration of uncertain but promising items.
- **Thompson Sampling**: Samples from a posterior distribution over reward probabilities for each item, balancing exploration probabilistically (Chapelle and Li 2011).

Use cases: News feed ranking, online ads selection, and quick product suggestions in low-latency environments.

5.1.2.2 Deep Q-Networks (DQN)

DQN extends *Q-learning*, a value-based reinforcement learning method, by using deep neural networks to approximate the *Q-function* $Q(s, a)$, which estimates the expected cumulative reward of taking action a in state s and following the policy thereafter (Mnih et al. 2015). Traditional Q-learning uses a tabular form, which is infeasible for large or continuous state-action spaces. DQN solves this by replacing the table with a neural network.

How it works: The agent selects actions by greedily maximizing the Q-values predicted by the network. During training, it minimizes the temporal difference loss between predicted and target Q-values. The target Q-value is computed using a separate target network to stabilize learning.

Key techniques:

- **Experience Replay**: Stores past transitions in a buffer and samples mini-batches randomly to break correlation between experiences.
- **Target Network**: Uses a slowly updated copy of the Q-network to compute target Q-values, reducing instability.

Use Cases: Well-suited for settings with high-dimensional states such as user profiles or interaction histories, including dynamic playlists, travel itinerary recommendation, or game-based item unlocking.

5.1.2.3 Policy Gradient Methods

Policy gradient methods directly optimize the parameters θ of a *stochastic policy* $\pi_\theta(a|s)$ to maximize the *expected return*:

$$ J(\theta) = \mathbb{E}_{\tau \sim \pi\theta} \left[\sum_t r_t \right] $$

How it works: These methods compute the gradient $\nabla_\theta J(\theta)$ using rollouts sampled from the policy. Unlike value-based methods like Q-learning, they do not require estimating a value function (though some variants do use a critic for variance reduction). The policy is updated through gradient ascent on the expected return.

Common algorithms:

- **REINFORCE**: A Monte Carlo method using full episodes to estimate the gradient (Williams, 1992).
- **Proximal Policy Optimization (PPO)**: A stable and sample-efficient algorithm that constrains updates via a clipped objective (Schulman et al., 2017).

Use Cases: Ideal for applications with large or structured action spaces, such as open-ended response generation, multi-step product configuration, or task-based conversational recommendation.

5.1.2.4 Monte Carlo Tree Search (MCTS)

MCTS is a *model-based planning algorithm* that constructs a *search tree* over possible future actions and outcomes, enabling lookahead-based decision-making (Brown et al. 2012).

How it works: MCTS iteratively builds a tree using four phases:

1. **Selection**: Traverse the tree from the root using a policy like UCB to balance exploration and exploitation.
2. **Expansion**: Add one or more child nodes to expand the tree.
3. **Simulation**: Run a rollout (e.g., random or policy-based) from the new node to estimate the outcome.
4. **Backpropagation**: Update value estimates of nodes along the path using the simulation result.

Value functions may be estimated via Monte Carlo averages or learned predictors. This method was famously used in AlphaGo (Silver et al., 2016) to combine neural value estimation with search-based planning.

Use Cases: Effective for long-horizon planning tasks in recommendation, such as curriculum sequencing, narrative arc planning, or accessory bundling (e.g., camera → lens → tripod).

In Table 5.1, we compare the aforementioned RL models in terms of their key features, and what each model is best suited for.

Table 5.1 Comparison of RL models for conversational recommendation systems

RL model	Key feature	Best suited for
Multi-armed bandit	Simple, fast adaptation, exploration-exploitation	Real-time product or news recommendation
Deep Q-network (DQN)	Deep Q-learning, state-action value estimation	Multi-turn dialogue systems, evolving preference modeling
Policy gradient methods	Direct policy optimization, flexible action space	Personalized conversational recommendation
Monte Carlo tree search	Sequential planning and simulation	Multi-step decision-making (e.g., cart-building recommendations)

5.1.3 Integrating RL with LLMs in Conversational Recommendation

Integrating Reinforcement Learning (RL) with Large Language Models (LLMs) enables conversational recommender systems that are both semantically fluent and behaviorally adaptive. LLMs bring strengths in natural language understanding, generation, and context tracking, while RL provides mechanisms to optimize decision-making based on long-term user feedback.

5.1.3.1 Roles of LLM and RL

- **LLM Component**: Responsible for interacting with the user via natural language, understanding user intent and generating fluent responses or recommendations (He et al. 2023).
- **RL Component**: Optimizes dialogue strategy by learning policies that maximize long-term rewards, such as user satisfaction, engagement, or diversity of exposure.

Example: In a fashion shopping assistant, the LLM generates outfit suggestions while the RL policy determines whether to continue suggesting, ask clarifying questions, or end the session, aiming to maximize session-level engagement.

5.1.3.2 Dialogue-Level Reward Design

RL facilitates the definition and optimization of reward functions at the dialogue level. Rewards may correspond to:

- Task success (e.g., item purchased or accepted)
- User satisfaction (e.g., feedback, sentiment, or dwell time)
- Diversity and novelty in recommendations (Christakopoulou et al. 2018)

5.1.3.3 Pipeline Integration Strategy

- **Stage 1**: Supervised fine-tuning of the LLM to learn task-specific dialogue flows using historical data.
- **Stage 2**: RL-based policy tuning on top of the fine-tuned LLM using either real user feedback or simulated interactions (Jaques et al. 2016).

5.1.3.4 Benefits of LLM-RL Integration

- **Adaptive Recommendations**: RL enables dynamic adaptation to user preferences during multi-turn dialogues.
- **Optimized Interaction Flow**: RL can adjust the sequencing of recommendation and clarification to optimize conversation outcomes.
- **Continuous Learning**: Policies evolve over time with more interactions, leading to better personalization and user retention.

This integration of RL and LLMs blends deep language understanding with adaptive policy optimization to create truly intelligent, conversational recommenders.

5.2 Key Modules in CRS

Conversational recommendation systems combine traditional recommendation techniques with real-time, dialogue-based interaction, offering a more intuitive and personalized experience. Unlike static recommenders, which rely on pre-computed suggestions, these systems actively engage users through dynamic conversations, clarifying preferences and adapting recommendations in real time. This interactive approach is particularly valuable in domains like e-commerce, entertainment, and travel, where user needs can evolve during the interaction.

Conversational recommendation systems (CRS) integrate multiple interdependent components to effectively manage dialogue, interpret user intent, personalize suggestions, and adapt over time (Sun & Zhang 2018). Each component plays a specific role in enabling the system to understand natural language inputs, deliver relevant recommendations, and continuously improve through feedback. Figure 5.1 presents the key modules of a conversational recommendation system. Table 5.2 provides a structured overview of these core modules and their associated techniques.

5.2 Key Modules in CRS

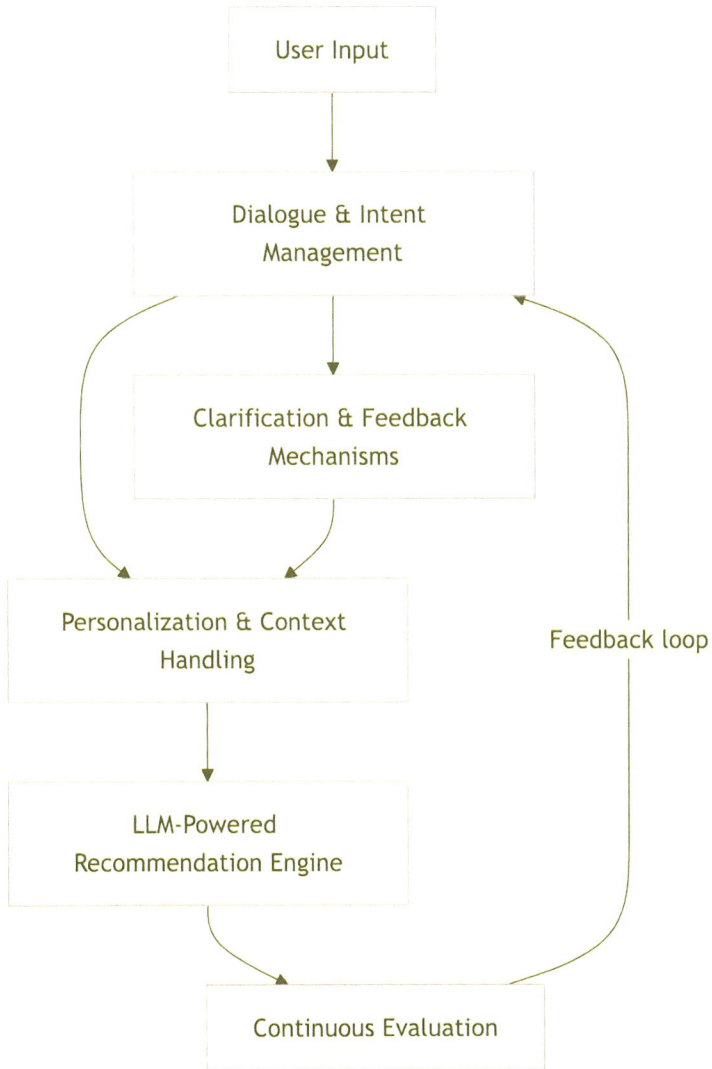

Fig. 5.1 Key Modules in Conversational Recommendation System

5.2.1 Dialogue and Intent Management

Effective dialogue and intent management are essential for task-oriented conversational recommender systems. These systems must understand and respond accurately to user requests over multiple interactions. This involves three tightly integrated tasks: *intent detection, slot filling*, and *dialogue state tracking*. Together, they enable the system to interpret evolving user inputs, extract meaningful details,

Table 5.2 Core components of conversational recommendation systems

Key components	Description	Key techniques used	Modules
Dialogue and intent management	Manages the flow of the conversation, tracks user intent, and extracts relevant information (slots) to guide recommendations	State tracking, NLU, NER, slot filling, pre-trained LLMs	– Intent detection–Slot filling–Context tracking
Clarification and feedback mechanisms	Handles clarification questions and adjusts recommendations based on user feedback and misinterpretations	Reinforcement learning (RL), active learning, contextual follow-up	– Clarification queries—Feedback loops—Error correction
Personalization and context handling	Customizes recommendations based on user preferences, historical data, and real-time context	User profiling, contextual embeddings, dynamic recommendation models	– User profiling—Context-aware recommendations—Real-time data adaptation
Continuous evaluation	Continuously assesses the effectiveness of recommendations based on user interactions and feedback	A/B testing, user satisfaction metrics, real-time performance tracking	– Model performance tracking—Iterative updates—Metrics analysis

and maintain contextual awareness across the dialogue (Henderson et al. 2014; Bordes and Weston 2017; Chen et al. 2019).

5.2.1.1 Intent Detection

What It Is

Intent detection is the process of identifying the user's goal or intention behind an utterance. In a recommendation scenario, this could include actions such as requesting a recommendation, filtering previous results, or asking for product specifications.

Why It's Needed

Accurate intent classification helps ensure the system responds appropriately. For example, the utterance "Show me affordable laptops" implies a price-filtering intent.

5.2 Key Modules in CRS

Techniques Used

Large Language Models (LLMs) like BERT or GPT are used for intent classification. These models can be:

- Fine-tuned on labeled intent data.
- Used in zero-shot settings with prompt engineering.

Example: Zero-Shot Intent Classification with OpenAI or Hugging Face

```
from transformers import pipeline
classifier = pipeline("zero-shot-classification",
model="facebook/bart-large-mnli")
utterance = "I'm looking for noise-cancelling headphones
under $200"
candidate_labels = ["get recommendation", "filter by price",
"ask for product specs", "request review"]
result = classifier(utterance, candidate_labels)
print("Predicted Intent:", result["labels"][0])
```

5.2.1.2 Slot Filling

What It Is

Slot filling involves extracting structured details, like category, budget, or brand, from user utterances. For example, from:

```
"I want a red dress under $100 for summer" The slots could
be: {"category": "dress", "color": "red", "price": "<100",
"season": "summer"}.
```

Why It's Needed

Slot filling enables fine-grained personalization and relevance in responses.

Techniques Used

Pre-trained models like BERT can be fine-tuned for Named Entity Recognition (NER) or slot tagging, using token classification tasks.

Example: Slot Tagging with BERT (Token Classification)

```
from transformers import AutoTokenizer,
AutoModelForTokenClassification
from transformers import pipeline
model_name = "dslim/bert-base-NER"
nlp = pipeline("ner", model=model_name, tokenizer=model_
name, grouped_entities=True)
text = "Find me a smartphone with good camera under $500"
slots = nlp(text)
# Example output (simplified):
# [{'entity_group': 'PRODUCT', 'word': 'smartphone'},
{'entity_group': 'FEATURE', 'word': 'camera'}, {'entity_
group': 'PRICE', 'word': '$500'}]
print("Extracted Slots:", slots)
```

5.2.1.3 Dialogue State Tracking (DST)

What It Is

Dialogue State Tracking maintains a dynamic summary of the conversation, such as inferred intent, filled slots, and unanswered questions.

Why It's Needed

It allows the system to respond coherently across multiple turns. Without DST, the system may forget user preferences, repeat itself, or offer irrelevant options.

Techniques Used

DST can be implemented as:

- A *slot-value memory structure* updated over each turn.
- A *prompt-based LLM* that tracks state implicitly.
- A *fine-tuned transformer* trained to output a JSON-style dialogue state.

Example: Simplified DST with Explicit State Updates

```
# Simulate maintaining a dialogue state
dialogue_state = {
    "intent": None,
    "slots": {}
}
# Example interaction
user_input_1 = "I'm looking for wireless headphones"
dialogue_state["intent"] = "get recommendation"
dialogue_state["slots"]["category"] = "headphones"
dialogue_state["slots"]["feature"] = "wireless"
user_input_2 = "Show me ones under $100"
dialogue_state["slots"]["price"] = "<100"
print("Current Dialogue State:", dialogue_state)
# Output: {'intent': 'get recommendation', 'slots':
{'category': 'headphones', 'feature': 'wireless', 'price':
'<100'}}
```

5.2.2 Clarification and Feedback Mechanisms

Clarification queries and feedback loops are critical components in LLM-powered conversational recommender systems (CRS). These mechanisms enable the system not only to generate relevant recommendations but also to dynamically adapt and improve its responses over time. By resolving ambiguous inputs and incorporating continuous user feedback, the system enhances personalization, strengthens user trust, and refines its understanding of user preferences (Li et al. 2016; Zhao et al. 2019; Christakopoulou et al. 2018).

- **Clarification Queries**: To resolve vague or ambiguous inputs, CRS systems generate follow-up questions like "Are you looking for wireless or noise-cancelling features?" This improves recommendation accuracy and mirrors natural dialogue (Zhao et al. 2019). Powered by LLMs with Chain-of-Thought reasoning and prompt engineering, these queries decompose user goals and identify missing information in multi-turn conversations.
- **Feedback Loops**: User responses provide real-time signals that help the system refine its recommendations and adapt to evolving preferences (Christakopoulou et al. 2018). Techniques such as Multi-Armed Bandits, DQN, and RLHF allow systems to optimize long-term engagement and overcome challenges like cold-start or preference drift (Li et al. 2016).

5.2.3 *Personalization and Context Handling*

Personalization in conversational recommender systems (CRS) is key to delivering tailored user experiences by leveraging individual preferences, historical.

5.2.3.1 Context Handling

- **What it is**: Context handling refers to the system's ability to interpret and adapt to evolving user behavior, dialogue history, and situational factors such as time, location, or device. This enables the recommender to stay responsive throughout the conversation.
- **Why it matters**: Without context awareness, the system may deliver stale or irrelevant results. Capturing context ensures continuity and relevance, especially in multi-turn dialogues where user preferences can shift dynamically (Zhao et al. 2019).
- **Key Techniques**:
    - **Dynamic embeddings**: Update user representation based on recent utterances.
    - **Hybrid models**: Integrate collaborative filtering, content, and session signals.

Example: Updating User embedding with New Context

```
# Initial user embedding (e.g., from past behavior)
user_embedding = np.array([0.3, 0.4, 0.2])
# Incorporate new input: "I'm looking for something lighter"
from transformers import AutoModel, AutoTokenizer
tokenizer
= AutoTokenizer.from_pretrained("bert-base-uncased")
model = AutoModel.from_pretrained("bert-base-uncased")
inputs = tokenizer("I'm looking for something lighter",
return_tensors="pt")
outputs = model(**inputs)
new_context_embedding = outputs.last_hidden_state.
mean(dim=1).detach().numpy()
# Blend with prior context
user_embedding = 0.7 * user_embedding + 0.3 * new_context_
embedding
```

5.2.3.2 Personalization

- **What it is**: Personalization tailors the recommendation journey to individual users by considering both immediate needs and long-term preferences. The system goes beyond reactive suggestions and proactively guides the user through a

coherent, personalized experience.
- **Why it matters**: Traditional recommenders often treat each request in isolation. Personalization creates continuity, for example, suggesting not just a movie, but trailers, reviews, and related content that match the user's taste (Zhang et al. 2018).
- **Techniques Used**:
    - **Reinforcement Learning (RL)**: Learns optimal policies for multi-turn interactions.
    - **LLM planning**: Uses Chain-of-Thought reasoning to anticipate future needs.

Example: Planning a Personalized Content Flow with LLM

```
import openai
prompt = """
User: I'm looking for a romantic movie to watch tonight.
System: Plan the next three steps to personalize the
conversation.
Output as JSON.
"""
response = openai.ChatCompletion.create(
    model="gpt-4",
    messages=[{"role": "user", "content": prompt}]
)
print(response['choices'][0]['message']['content'])
```

5.2.4 Continuous Evaluation

Conversational recommendation systems (CRS) operate in dynamic environments where user preferences and interaction patterns change over time. Continuous evaluation ensures that these systems remain effective, adaptive, and aligned with evolving user expectations. It allows developers to identify issues early, refine system behavior, and sustain long-term performance.

5.2.4.1 Evaluation Data Sources

- **Wizard-of-Oz (WOZ) Simulations**: Human evaluators simulate system responses to assess naturalness, satisfaction, and task completion in controlled scenarios.
- **Multi-Turn Dialogue Datasets**: Benchmarks such as *DSTC* and *MultiWOZ* capture realistic user-system conversations across domains and provide test beds for dialogue modeling and context tracking.

5.2.4.2 Evaluation Metrics

- **Task Completion**: Measures success in helping users achieve goals (e.g., purchase, booking, selection).
- **Natural Language Generation (NLG) Metrics**:
  - *Relevance:* Response aligns with user intent.
  - *Coherence:* Logical and context-consistent dialogue.
  - *Engagement:* Conversational flow is interesting and natural.
- **Statistical Metrics**: BLEU, ROUGE (n-gram overlap); Perplexity (language fluency).
- **Behavioral Metrics**: Implicit signals such as click-through rate, dwell time, or user ratings.

Regular evaluation ensures iterative improvement and robust adaptation to user needs, keeping the CRS system effective and engaging in real-world deployments.

5.2.5 Reward Design in CRS

- **What it is**: Reward design specifies what outcomes the system should optimize for, for example, user satisfaction, task completion, or engagement.
- **Why it matters**: With well-defined rewards, agents can learn not only to recommend but to clarify, explore diverse options, and sustain long-term user engagement (Zhao et al. 2019; Christakopoulou et al. 2018).
- **Common Reward Signals**:
  - **Task success**: User accepts or purchases an item.
  - **User satisfaction**: Inferred from clicks, sentiment, dwell time.
  - **Diversity**: Rewards novel or serendipitous suggestions.

5.2.5.1 Example: Reward Computation for a Dialogue Turn

```
def compute_reward(user_action, recommendation,
dialogue_turns):
    reward = 0
    if user_action == "accepted":
        reward += 1.0
    if "thank you" in dialogue_turns[-1].lower():
        reward += 0.5
    if recommendation not in previous_recommendations:
        reward += 0.3  # diversity bonus
    return reward
```

5.3 Designing Conversational Recommender Systems

5.3.1 System Architecture and Workflow Integration

To effectively deploy LLM-powered Conversational Recommender Systems (CRS), it is essential to adopt a modular system architecture that supports real-time interaction, personalization, and scalability. This section outlines a practical design workflow, infrastructure needs, and performance optimization strategies.

Figure 5.2 illustrates the high-level system architecture of a conversational recommendation system, highlighting key functional blocks and their interactions:

- User input is processed by the Dialogue Management module, which handles intent understanding, dialogue state tracking, and policy decisions.
- The Context Handler enriches this interaction by incorporating user profiles, session data, and preferences.
- The Recommendation Engine generates tailored responses using retrieval, ranking, and LLM-based generation techniques, drawing from product and knowledge databases.
- Outputs are presented to the user via natural language, and the Feedback Loop captures interaction signals for continuous improvement, optionally supporting online model adaptation.

5.3.2 Data and Infrastructure Requirements

Conversational recommendation systems rely on several core data and infrastructure components to operate effectively:

- **User Data and Context Signals**: Include user profiles, interaction history, and real-time contextual cues (e.g., session behavior, preferences), powering the Context Handler for dynamic personalization.
- **Dialogue Logs and Feedback Data**: Collected through the Feedback Loop, these logs support supervised fine-tuning, reinforcement learning, and ongoing system evaluation.
- **Item and Knowledge Repositories**: Serve as the foundation for the Recommendation Engine, providing structured metadata, embeddings, and domain knowledge used in retrieval and generation.
- **Model Orchestration Infrastructure**: Manages the coordination between dialogue management, LLM inference, and recommendation workflows across components.
- **Scalable Deployment (Cloud/Edge)**: Ensures low-latency responses and system scalability, supporting deployment of LLMs and backend services at production scale.

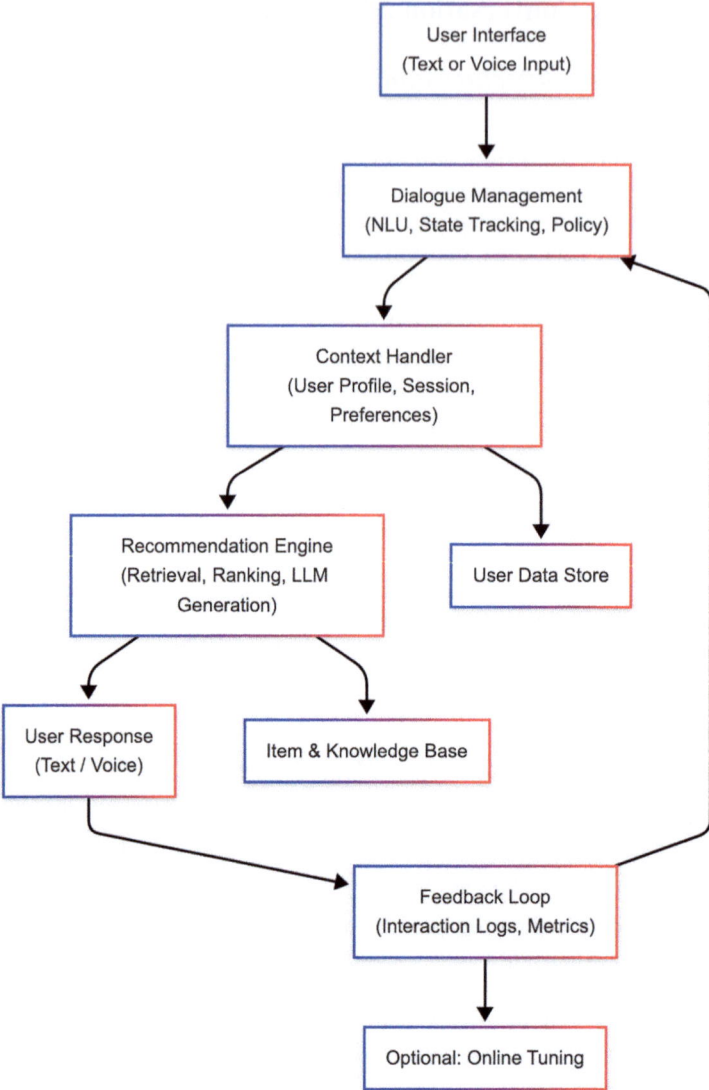

Fig. 5.2 System Architecture of a Conversational Recommendation System (CRS)

5.3.3 *Performance Optimization and Iterative Improvement*

Deploying a conversational recommendation system (CRS) in real-world settings requires careful consideration of both performance optimization and iterative improvement. This section discusses practical techniques to ensure responsiveness, cost-efficiency, and continuous system refinement based on user behavior.

5.3 Designing Conversational Recommender Systems

5.3.3.1 System Optimization Techniques

To meet production-level performance and scalability demands, CRS systems often employ the following strategies:

- **Caching and Response Reuse**:
  - Cache LLM responses for frequently asked queries or common intents.
  - Serve pre-computed recommendations for cold-start users to reduce latency.

- **Model Distillation and Lightweight Alternatives**:
  - Use distilled or quantized versions of large LLMs for routine tasks.
  - Implement model routing: Assign smaller models to lightweight interactions, reserving full LLMs for complex queries.

- **Latency-Aware Pipeline Design**:
  - Parallelize tasks such as intent recognition and retrieval.
  - Preload likely response templates while the user is typing to improve perceived responsiveness.

- **Fallback Strategies**:
  - Use rule-based templates when the LLM fails, times out, or generates invalid output.
  - Hybrid systems can blend heuristic outputs with LLM-generated content to ensure robustness and continuity.

These optimization mechanisms help balance system quality, cost, and reliability, particularly important in high-throughput environments such as e-commerce, streaming platforms, or customer service.

5.3.3.2 Performance Tracking and Iterative Refinement

Beyond initial deployment, sustained system performance depends on rigorous monitoring and continuous improvement. Key mechanisms include:

- **Real-Time Performance Monitoring**: Track key indicators such as response latency, user engagement, click-through rates, and satisfaction scores during live interactions.
- **A/B Testing and Controlled Experiments**: Evaluate system enhancements by comparing different model versions or interaction strategies. Metrics such as task completion, dwell time, and user ratings help assess impact objectively.
- **Dynamic Model Updates**: Incorporate fresh interaction logs and retrain or fine-tune models regularly to adapt to evolving user behavior or content trends.
- **Feedback Loop Integration**: Utilize user feedback to refine both LLM prompts and RL reward functions. Iteratively improving models ensures long-term system quality and personalization.

5.4 Tutorial: Conversational Recommendation System with RL and LLMs

5.4.1 Overview

This tutorial provides a hands-on example of how RL and LLMs can work together to create personalized, interactive recommendation systems. We use the example of purchasing headphones through a conversational chatbot. By simulating user interactions, we show how LLMs can be used for extracting preferences from natural language, while RL optimizes recommendations based on learned rewards.

Goal of the Tutorial
- *Understand how to extract user preference* from dialogue history using LLMs.
- *Familiar with RL reward design* to balance user constraints and satisfaction.
- *Experience with end-to-end integration* of conversational AI and decision-making systems.

We show a condensed version of this tutorial in the book text. The full version of the code is available at: https://github.com/qqwjq1981/springer-LLM-recommendation-system

5.4.2 Experimental Design

5.4.2.1 Dataset Design

- **Simulate User Interaction History**: Synthetic dialogs with explicit/implicit preferences (e.g., "Over-ear under $100").

```
Example:
conversation = [
    {"user": "Hi, I'm looking for wireless headphones.",
"bot": "Do you prefer over-ear or in-ear?"},
    {"user": "Over-ear, under $100.", "bot": "Recommended:
Sony WH-CH510 ($80). More options?"},
    {"user": "Yes.", "bot": "JBL Tune 510BT ($100)."},
    {"user": "I'll take Sony.", "bot": "Great choice!"}
]
```

- **User query**:

```
"I need comfortable headphones for travel with good noise
cancellation. My budget is around $100."
```

5.4 Tutorial: Conversational Recommendation System with RL and LLMs   149

- **Product Catalog**: Collection of real headphones from Amazon, tagged with title, product type, features, brand and price, to align with user preference features.

```
{'product_title': 'FIGMASU Headphones Wireless Bluetooth
Neckband Wireless Headsets for Sport',
'product_type': 'on-ear',
'features': ['built-in microphone', '100 H playtime',
'sweatproof'],
'brand': 'FIGMASU',
'price': 29.98}
```

5.4.2.2 Methodology

- **User Preference Extraction**: We used LLM to extract user preference from conversation history and current query, and then merge to receive final inferred user preference.

```
# Interaction History Processor Prompt
{"role": "system", "content": """Analyze conversation
history and respond with a JSON object:
- preferred_brand: string
- avoided_features: list
- budget_range: [min,max]
- implicit_type_pref: string"""},
{"role": "user", "content": history_str}
# Query Intent Classification Prompt
{"role": "system", "content": """Extract EXACT preferences
from this query as JSON with:
- type: over-ear/in-ear/earbuds
- price_max: number
- features: list
- use_case: string
- urgency: high/medium/low"""},
{"role": "user", "content": query}
```

- **RL Environment**: Simulates a recommendation space with products and user feedback. Table 5.3 summarizes the mapping of the headphone recommendation example to standard RL terminology.
  - **States**: User preferences + current intent.
  - **Actions**: Which product to recommend.
  - **Rewards**: we consider two reward functions, one is weighted version of number of matches between user preference and product features (Feature Match

Table 5.3 Mapping the headphone recommendation example to RL concepts

RL concept	Recommendation system equivalent	Code reference
Agent	Recommendation bot	PPO("MlpPolicy", env)
Environment	Simulated user interaction space	RecommendationEnv(products, prefs)
State	User preferences + current intent	env.state
Action	Product recommendation	model.predict(state) → product index
Reward	User acceptance (+1) or rejection (−1)	env.step(action) returns reward
Policy	PPO's neural network decision-making rules	model.policy

Reward) and another is feature match reward with a piecewise linear component to reflect price preference (Piecewise Linear Reward).

Feature match reward:

```
feature_score = len(user_features & product_features) /
max(1, len(user_features))
# Type and brand matching
type_match = product['product_type'] == self.user_
preferences.get('product_type', '')
brand_match = product['brand'].lower() == self.user_
preferences.get('brand', '').lower()
price_budget = self.user_preferences.get('price_max',
float('inf'))
if self.safe_price(product.get('price', 1000)) >
price_budget:
return 0.0
return min(1.0, 0.8 * type_match + 0.2 * brand_match)
```

Piecewise linear reward function: That considers user favor of premium features and places overshoot penalty.

```
price_ratio = min(1.0, price / price_budget) if price_budget
> 0 else 0
# Apply price scaling
if price <= price_budget:
    price_modifier = 0.5 * (1 + price_ratio)  #
0.5-1.0 scaling
else:
    overshoot = (price - price_budget) / max(1,
price_budget)
    price_modifier = max(0, 0.5 - 0.2 * overshoot)  #
Penalize overspending
```

- **RL Model**:

  Proximal Policy Optimization (PPO) trains a policy to maximize cumulative rewards.

5.4.3 Results and Analysis

5.4.3.1 User Preference Extraction

- **Interaction History Processor** successfully inferred preferences.

```
{'preferred_brand': 'Sony',
 'avoided_features': ['earbuds'],
 'budget_range': [0, 100],
 'implicit_type_pref': 'over-ear'}
```

- **Intent Classification** accurately classified intents (e.g., "budget_constraint" for *"under $100"*).

```
{'type': 'over-ear',
 'price_max': 100,
 'features': ['comfortable', 'good noise cancellation'],
 'use_case': 'travel',
 'urgency': 'medium'}
```

- **Merged User Preference**: we adopted simple logic to merge the preference inferred from previous conversations and current query.

```
{'product_type': 'over-ear',
 'price_max': 100,
 'features': ['comfortable', 'long-battery', 'good noise cancellation'],
 'brand': 'Sony'}
```

5.4.3.2 Reward Dynamics

We examined RL training with 50 episodes and 200 steps in each episode, and examined total reward over time during the training period and test results.

- The reward converges much faster with a discrete reward computed based on the number of feature matches, but much slower when the reward is more continuous with the addition of non-trivial price preference.

Table 5.4 Reward dynamic over training and test results, under Feature Match Reward and Piecewise Linear Reward

Reward	Feature Match Reward	Piecewise Linear Reward
Reward dynamic	Reward Over Training (feature_count rewards) — total reward rises quickly to ~4.0 over ~50 episodes	Reward Over Training (piecewise_linear rewards) — total reward rises quickly to ~7.0 over ~50 episodes
Model response	Step 1: OneOdio A11 Wireless Headphones Over Ear | Type: over-ear | Price: $19.99 | Reward: 0.80 Step 2: Sleep Headphones, Wireless Music Eye Mask | Type: over-ear | Price: $22.99 | Reward: 0.80 Step 3: Manhattan Stereo Headset | Type: over-ear | Price: $9.99 | Reward: 0.80 Step 4: Sleep Headphones, Wireless Music Eye Mask | Type: over-ear | Price: $22.99 | Reward: 0.80 Step 5: Sleep Headphones, Wireless Music Eye Mask | Type: over-ear | Price: $22.99 | Reward: 0.80	Step 1: Earpads Compatible with LS31 LS41 LS35X LS50X Headset with Microphone Foam I Replacement Ear Cushion (Cooling Gel Fabric) | Type: over-ear | Price: $29.99 | Reward: 1.45 Step 2: Earpads Compatible with LS31 LS41 LS35X LS50X Headset with Microphone Foam I Replacement Ear Cushion (Cooling Gel Fabric) | Type: over-ear | Price: $29.99 | Reward: 1.45 Step 3: Earpads Compatible with LS31 LS41 LS35X LS50X Headset with Microphone Foam I Replacement Ear Cushion (Cooling Gel Fabric) | Type: over-ear | Price: $29.99 | Reward: 1.45 Step 4: Earpads Compatible with LS31 LS41 LS35X LS50X Headset with Microphone Foam I Replacement Ear Cushion (Cooling Gel Fabric) | Type: over-ear | Price: $29.99 | Reward: 1.45 Step 5: Sleep Headphones, Wireless Music Eye Mask | Type: over-ear | Price: $22.99 | Reward: 1.41

- Both reward functions lead to a diverse set of recommendation results. However, the Piecewise Linear Reward tends to lead to premium product recommendations due to the premium price preference in the reward function (Table 5.4).

5.4.3.3 Discussions

1. **Effectiveness of RL**:
   - The RL model successfully learns to recommend products that align with user preferences, as evidenced by the increasing frequency of positive feedback.

2. **Role of LLMs**:
   - LLMs play an important role in extracting user preferences from conversational data, enabling personalized recommendations.
3. **Limitations**:
   - The model's performance depends on the quality of the reward function and the diversity of the product catalog.
   - More training data and longer training times may be required for complex scenarios.
4. **Future directions**: Future direction on the study could include:
   - **More sophisticated reward functions**: Additional rewards can include a *diversity bonus* to encourage varied recommendations, a *dialogue efficiency reward* favoring shorter successful conversations, and a *confirmation reward* based on positive user responses like "Yes, that works."
   - **Fine-tuning LLMs**: Fine-tune the LLM on domain-specific data to improve preference extraction accuracy.
   - **Multi-Objective RL**: Optimize for additional objectives, such as diversity of recommendations or long-term user engagement.

5.5 Conclusion

This tutorial demonstrates how RL and LLMs can be combined to build a conversational recommendation system. By extracting user preferences and optimizing recommendations through RL, the system achieves personalized and interactive recommendations. The results highlight the potential of this approach for real-world applications, while also identifying areas for future improvement.

References

Bordes, A., & Weston, J. (2017). *Learning end-to-end goal-oriented dialog*. ICLR.

Brown, C. B., Powley, E., Whitehouse, D., et al. (2012). A survey of Monte Carlo tree search methods. *IEEE Transactions on Computational Intelligence and AI in Games, 4*(1), 1–43.

Chapelle, O., & Li, L. (2011). *An empirical evaluation of Thompson sampling*. NeurIPS.

Chen, Q., Zhuo, Z., & Wang, W. (2019). *BERT for joint intent classification and slot filling*. arXiv:1902.10909.

Christakopoulou, K., Beutel, A., Covington, P., & Lefakis, L. (2018). *Q&R: A two-stage approach toward interactive recommendation*. Proceedings of the 12th ACM Conference on Recommender Systems.

He, J., Zhang, M., Liu, Y., et al. (2023). *Large language models as zero-shot conversational recommenders*. arXiv:2308.10053.

Henderson, M., Thomson, B., & Williams, J. D. (2014). *The second dialog state tracking challenge*. SIGDIAL.

Jaques, N., Gu, S., Bahdanau, D., et al. (2016). *Sequence tutor: Conservative fine-tuning of sequence generation models with KL-control*. ICML.

Lattimore, T., & Szepesvári, C. (2020). *Bandit algorithms*. Cambridge University Press.

Li, J., Monroe, W., Shi, T., et al. (2016). *Deep reinforcement learning for dialogue generation*. EMNLP.

Mnih, V., Kavukcuoglu, K., Silver, D., et al. (2015). Human-level control through deep reinforcement learning. *Nature, 518*(7540), 529–533.

Schulman, J., Wolski, F., Dhariwal, P., Radford, A., & Klimov, O. (2017). *Proximal policy optimization algorithms*. arXiv:1707.06347.

Silver, D., Huang, A., Maddison, C. J., et al. (2016). Mastering the game of go with deep neural networks and tree search. *Nature, 529*(7587), 484–489.

Sun, Y. and Zhang, Y. (2018). *Conversational recommender system*. SIGIR.

Sutton, R. S., & Barto, A. G. (2018). *Reinforcement learning: An introduction* (2nd ed.). MIT Press.

Williams, R. J. (1992). Simple statistical gradient-following algorithms for connectionist reinforcement learning. *Machine Learning, 8*(3), 229–256.

Zhang, Y., Gong, Y., Wu, Q., et al. (2018). *Towards personalized conversational recommendation*. RecSys.

Zhao, S., Zhao, D., & Eskenazi, M. (2019). *Improving user satisfaction with clarification questions*. NAACL.

Chapter 6
Leveraging Multi-modal Data

This chapter examines how multi-modal data like text, image, audio, and videos can enhance recommendation systems. It introduces core integration strategies (early, late, and hybrid fusion) and contrasts them with emerging multi-modal large language models (LLMs) in terms of architecture, training, and use cases.

A practical tutorial on fashion recommendation using the Amazon Fashion dataset demonstrates how CLIP embeddings can be used in a pairwise ranking task. Experimental results compare a neural MLP-based model with a dot-product baseline, highlighting the benefits and trade-offs of learning non-linear user preferences from multi-modal inputs.

6.1 Introduction

Modern recommendation systems must grapple with increasingly complex user preferences and multifaceted content. While traditional text-based LLMs excel at parsing linguistic patterns, they lack the ability to interpret the visual, auditory, and behavioral cues that define user intent in domains like e-commerce, entertainment, and conversational interfaces. This gap has driven the evolution of LLMs toward *multi-modal architectures* that unify text, images, audio, and video into a cohesive understanding framework (Lu et al., 2019; Wei et al., 2024).

6.1.1 Core Modalities and Their Roles

Table 6.1 summarizes the core modalities prevalent in recommendation systems:

Table 6.1 Commonly used data modalities (text, images, audio, and video) and their roles in recommendation systems

Modality	Key data sources	Role in recommendations	Example use case
Text	Product descriptions, reviews, search queries	Captures semantic preferences (e.g., "organic skincare"), sentiment analysis, and intent parsing	Matching "spicy fragrances" to perfumes with pepper/cinnamon notes
Images	Product photos, thumbnails, user-generated content	Infers aesthetic preferences (color, style) and visual similarity	Recommending handbags with similar shapes/colors to pinned items
Audio	Voice queries, music tracks, podcasts	Identifies acoustic preferences (tempo, mood) and vocal tone in interactions	Suggesting "upbeat acoustic covers" based on liked songs
Video	Product demos, short-form content	Analyzes temporal engagement (rewatched segments) and contextual behavior	Recommending DIY tools after watching home renovation tutorials

- **Text**: Textual data include product descriptions, user reviews, titles and keywords, and they help the system understand the meaning, sentiment, and attributes of an item or user.
- **Images**: Visual data (e.g., product images or movie posters) provides important context about the look and style of the item, influencing user preferences.
- **Audio**: In applications like conversational agents, audio data can help capture tone, intent, and emotion, providing deeper insights into user preferences.
- **Video**: Video content such as product demos, movie trailers, live streams, or user-generated clips offers rich temporal and multi-sensory information that can capture dynamic aspects of an item (e.g., fit, usage, atmosphere). In recommendation systems, video helps infer *style, pacing, emotional tone, and functional attributes*, often combining visual, audio, and textual cues.

6.1.2 The Multi-modal Advantage

Integrating multi-modal data allows recommendation systems to better capture user intent and content features, especially in complex, context-driven domains.

- **Enriching User Profiles**: Combining reviews, queries, and browsing history with visual or audio cues (e.g., product images or sound clips) enables more nuanced modeling of preferences, such as identifying aesthetic taste from clicked images.
- **Improving Content Understanding**: Multi-modal signals enhance how systems represent and compare content. For example, a movie recommender might combine user review sentiment with poster imagery to align emotional tone and visual appeal.

- **Supporting Multi-modal Interaction**: Conversational systems benefit from multi-modal inputs—spoken requests, uploaded images, or mixed inputs—allowing richer, more flexible, and personalized real-time recommendations.

6.1.3 Challenges in Multi-modal Integration

While multi-modal data enriches user and item representations, it introduces several challenges in modeling and computations:

- **Modality Alignment**: Different modalities capture complementary aspects (e.g., text mentions color, image shows it). Misalignment between modalities can reduce learning effectiveness (Tsai et al., 2019).
- **Feature Extraction**: Each modality requires specialized models—BERT for text, CNNs for images—making it difficult to unify them without losing important details (Gao et al., 2020).
- **Representation Fusion**: Poor fusion strategies can lead to overfitting or one modality dominating. Balanced techniques like cross-attention or co-embedding are needed (Liu et al., 2023).
- **Computational Overhead**: Multi-modal models demand more processing and storage, requiring efficient design for scalability.

6.1.4 Modeling Strategies

In Sects. 6.2 and 6.3, we will explore how LLMs use multi-modal data to improve recommendation accuracy, the challenges involved in fusing different data types, and the techniques used for effective multi-modal integration.

- **Multi-modal integrations**: combine models or systems that specialize in different modalities (e.g., a text model, an image model) into a *coordinated pipeline* to handle multi-modal data. We can use multi-modal integration to transform the data into *shared embedding representation*.
- **Multi-modal LLMs**: End-to-end models that incorporate cross-modal attention to jointly learn from multiple input types. These models simplify architecture while maintaining or improving performance.

6.2 Multi-modal Integration Techniques

In a multi-modal integrated system, each modality is typically processed by a specialized model tailored to its data type. Common model choices include:

- **Text**: GPT-style Transformers or BERT variants for semantic understanding and intent parsing (Devlin et al., 2019)
- **Images**: Convolutional Neural Networks (e.g., ResNet) for extracting visual features like color, texture, and shape (He et al., 2016)
- **Audio**: CNN-RNN hybrids or spectrogram-based models (e.g., VGGish, YAMNet) for capturing acoustic patterns and vocal cues (Hershey et al., 2017)
- **Video**: 3D CNNs or Vision Transformers for modeling temporal and visual dynamics across frames (Arnab et al., 2021)

These models extract modality-specific features, which are then fused to support recommendation or content understanding. This *modular design* allows developers to leverage state-of-the-art models for each input type without training a unified multi-modal system from scratch (Baltrusaitis et al., 2019; Tsai et al., 2019). This plug-and-play architecture offers flexibility and scalability, making it easier to adapt multi-modal recommendation systems to different domains and deployment settings.

Multi-modal integrations are particularly useful when high-performing, domain-specific models are already available. The fusion of their outputs can significantly enhance system performance, especially in domains like e-commerce, entertainment, and conversational AI.

The integration could be done by merging features before passing them through a final model (early fusion) or combining results after individual model outputs (late fusion). There are three primary techniques used to fuse multi-modal data in recommendation systems: *early fusion, late fusion*, and *hybrid fusion*. Each technique has its advantages depending on the task and the nature of the data. Figure 6.1 graphically represents modality fusion paradigms.

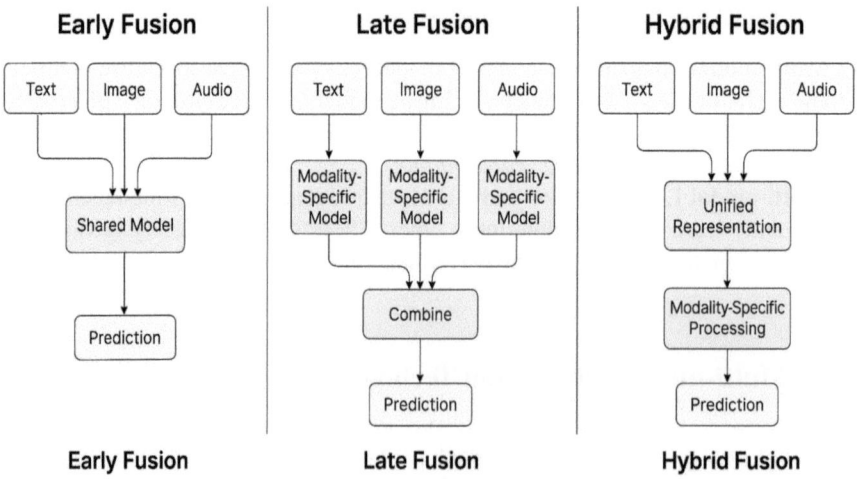

Fig. 6.1 Modal fusion paradigms

6.2.1 Early Fusion

In early fusion, *multiple modalities* (e.g., text, images, and audio) are combined before being input into the model. This technique *concatenates* or *averages* the embeddings from each modality, creating a single, unified representation that can then be used for further processing.

- **Advantages**: Early fusion provides a direct way of combining different data types into a single representation, which can be effective when the modalities are highly interdependent.
- **Example**: In an e-commerce recommendation system, an early fusion approach might concatenate the text embedding of a product description with the image embedding from a CNN to form a joint feature vector, which is then used to recommend products

```
# Example of early fusion (concatenating text and image
embeddings)
import torch
# Assume `text_embed` and `img_embed` are pre-trained
embeddings
# of text and image modalities
fused_embed = torch.cat((text_embed, img_embed), dim=1)
```

6.2.2 Late Fusion

In late fusion, each modality is processed independently, and the results are combined at the *decision-making* stage. This technique allows the model to maintain the integrity of each modality, processing them separately before merging the outputs for final recommendations.

- **Advantages**: Late fusion is less computationally intensive and offers more flexibility by allowing each modality to be treated with the most appropriate method (e.g., text with LLMs, images with CNNs).
- **Example**: In a movie recommendation system, late fusion might involve generating separate recommendations from text (reviews) and images (posters). The final recommendation is based on a weighted combination of the outputs from each modality.

```
# Example of Late Fusion (independently processing text
and image)
text_score = process_text(text_embed)
img_score = process_image(img_embed)
# Combine scores from text and image
final_score = 0.7 * text_score + 0.3 * img_score  # Weighted
combination
```

6.2.3 Hybrid Fusion

Hybrid fusion combines both early and late fusion techniques. Initially, different modalities are processed separately, but their embeddings are merged early in the process and then refined through independent processing steps. This allows the model to capture both intermodal relationships and preserve the individuality of each modality.

- **Advantages**: Hybrid fusion maximizes the strengths of both early and late fusion, allowing for nuanced representations and more robust performance, particularly when dealing with complex data.
- **Example**: In a music recommendation system, text data (e.g., lyrics) and audio features (e.g., tempo, pitch) might be first processed separately but combined to form a unified feature vector that is further refined by the model to make the final recommendation.

```
# Example of Hybrid Fusion (combining early and late fusion)
text_img_embed = torch.cat((text_embed, img_embed), dim=1)
# Early fusion
refined_embed = refine_embedding(text_img_embed)  # Further processing
final_score = process_with_other_modalities(refined_embed)  # Late fusion
```

6.3 Multi-modal LLMs

Multi-modal Large Language Models (LLMs) are designed to process and integrate information from diverse data modalities within a unified, end-to-end architecture. Unlike traditional multi-modal systems that rely on separate models per modality, multi-modal LLMs enable *joint learning* of cross-modal representations, allowing for richer modeling of complex interactions across modalities (Radford et al., 2021; Alayrac et al., 2022).

Multi-modal LLMs can be used for a range of tasks:

- Multi-modal retrieval (e.g., text-to-image search)
- Cross-modal reasoning (e.g., answering questions based on both text and images)
- Generative recommendations (e.g., explaining recommendations based on image and text inputs)
- Context-aware summarization or personalization using rich user-item representations

6.3.1 Modeling Principles of Multi-modal LLMs

Multi-modal LLMs are designed to process and integrate diverse input modalities such as text, images, audio, and video. The effectiveness of these models hinges on how they *encode, fuse*, and *align* information from different modalities. These three stages—*tokenization/encoding, fusion*, and *alignment*—form the backbone of multi-modal modeling and directly influence architectural choices.

6.3.1.1 Tokenization and Modality Encoding

The first step in multi-modal modeling is to convert each modality into a sequence of model-compatible representations:

- *Text* is typically tokenized using subword units (e.g., BPE, WordPiece).
- *Images* are broken down into patches (as in Vision Transformers) or represented through CNN features.
- *Audio* and *video* are segmented into frames or spectrograms.

To distinguish between modalities, models often append *modality-specific type embeddings* or *positional encodings*. This allows the model to condition processing based on source modality and preserve temporal/spatial structure.

6.3.1.2 Fusion Strategies

Fusion refers to how and when the model integrates information from different modalities. The major approaches include:

- **Single-stream fusion**: All modalities are concatenated and passed through a shared transformer. Cross-modal interactions emerge via self-attention (e.g., VisualBERT, VL-BERT). This approach allows tight integration but may struggle with modality imbalance.
- **Dual-stream fusion**: Separate encoders process each modality independently, followed by a *cross-attention* mechanism to align and exchange information (e.g., CLIP, LXMERT). This encourages strong modality-specific encoding and is particularly suited to retrieval or matching tasks.
- **Intermediate (hybrid) fusion**: Combines early modality-specific encoding with late shared transformer layers to integrate representations (e.g., Flamingo). This balances specialization with deep cross-modal reasoning.

6.3.1.3 Cross-Modal Alignment Objectives

To train these models, various objectives are employed to encourage alignment across modalities:

- **Contrastive learning**: Encourages matched text-image (or other modality) pairs to be close in the embedding space, while pushing apart mismatched pairs (e.g., CLIP).
- **Masked prediction**: Learns modality-specific understanding by predicting masked tokens or patches (e.g., BEiT-style objectives).
- **Cross-modal matching**: Trains models to identify whether inputs from different modalities correspond to the same instance.
- **Generative objectives**: Enable models to produce one modality conditioned on another (e.g., image-to-text or text-to-image generation).

These objectives can be used individually or in combination, depending on the target application.

6.3.1.4 From Principles to Model Designs

These encoding, fusion, and alignment strategies manifest in a spectrum of multi-modal LLM architectures. For example:

- **CLIP (Contrastive Language–Image Pre-training)**: A dual-stream model with independent encoders for images and text, trained with a contrastive loss to align them in a shared embedding space. Its design favors zero-shot retrieval and classification tasks (Radford et al., 2021).
- **BLIP/BLIP-2**: Flexible models combining dual encoders with a language decoder, enabling both contrastive pre-training and image-conditioned text generation.
- **Flamingo**: A hybrid model that processes modalities independently at first, then integrates them via a shared transformer block. It is optimized for few-shot visual reasoning tasks (Alayrac et al., 2022).
- **GPT-4V (GPT-4 with Vision)**: A unified multi-modal transformer that accepts both image and text tokens, enabling joint processing and generation. It builds on the single-stream principle, adapted for large-scale, general-purpose reasoning.

6.3.2 Advantages and Limitations

Advantages

- **Simplified Integration**: Single-model approach eliminates the need for separate pipelines.
- **Enhanced Understanding**: Deep integration of modalities improves context and personalization.
- **End-to-End Learning**: Joint optimization produces more coherent representations.

Limitations

- **Computational Complexity**: Increased complexity leads to higher resource requirements.
- **Reduced Interpretability**: End-to-end models make it harder to isolate the contribution of individual modalities.
- **Resource Balancing**: Ensuring equal representation across modalities can be challenging.

6.3.3 Choice Between Multi-modal Integrations and Multi-modal LLMs

Table 6.2 summarizes the differences between multi-modal LLMs and multi-modal integration in terms of architectures, data handling, and model training with examples.

Use Multi-modal Integrations if:

- You have well-established models for each modality (e.g., a strong image model like ResNet and a robust language model like GPT or BERT) and prefer integrating these models without significant retraining or modification.
- You favor a modular approach that allows leveraging the best available model for each modality and combining their outputs for the final task.
- The data or task does not demand extensive cross-modal interaction, meaning the system simply needs to aggregate insights from different domains.

Use Multi-modal LLMs if:

- Your application requires deep, end-to-end integration of diverse data types, where nuanced intermodal relationships can substantially enhance recommendation quality.
- You are tackling scenarios where complex, cross-modal interactions are critical, and the benefits of unified processing outweigh the higher computational costs.

Table 6.2 Differences between multi-modal LLMs and multi-modal integrations

Aspect	Multi-modal integrations	Multi-modal LLMs
Architecture	Combination of different specialized models for each modality	Single, unified model that processes multiple modalities
Data handling	Different models handle each modality, and outputs are *integrated*	Handles multiple types of data (text, images, audio, etc.) in a *single model*
Example	Text model (LLM) for reviews + image model (CNN) for product images, integrated for final recommendation	A single LLM model that processes both text (reviews) and images (product photos)
Training	Each modality can be trained independently, potentially using pre-trained models	Typically requires *joint training* on multiple modalities

- You have sufficient computational resources and can accept a potential reduction in interpretability in exchange for a streamlined integration pipeline that maximizes cross-modal learning.

6.4 Tutorial: Multi-modal Fashion Recommendation with Pairwise Ranking

6.4.1 Overview

This tutorial explores the construction of a *multi-modal recommendation system* using the Amazon Fashion dataset, which provides both *product images* and *textual descriptions*. The task is to learn user preferences and rank candidate items by relevance. We compare two models:

- *MLP-based neural ranking*, which captures *non-linear interactions* between user and item embeddings.
- *Dot-product similarity*, a *lightweight baseline* that assumes user preference is aligned with embedding proximity.

We use Bayesian Personalized Ranking (BPR) loss, which optimizes relative preference between positive and negative item pairs (e.g., "user prefers A over B"). This setup allows us to investigate trade-offs in model complexity, multi-modal fusion strategies, and common failure cases in the fashion domain, where visual style and personal taste can be subtle and subjective.

Goal of the Tutorial

- *Integrate multi-modal features* by combining CLIP-based text and image embeddings to represent items in a recommendation setting.
- *Compare ranking architectures*, evaluating the effectiveness of MLP-based models versus dot-product similarity for modeling user-item relevance.
- *Apply pairwise learning objectives* using BPR loss to optimize recommendations and understand trade-offs between model complexity, generalization, and interpretability in the fashion domain.

We show a condensed version of this tutorial in the book text. The full version of the code is available at: https://github.com/qqwjq1981/springer-LLM-recommendation-system

6.4.2 Experimental Design

Dataset and Preprocessing

- **Data Source**: Amazon Fashion dataset with aligned product images and text.
- **Splitting Strategy**: 80/20 user-wise split to prevent information leakage.
- **Item Embeddings**: Combined CLIP text and image features (averaged into a 768-dimensional vector).

```
# Load model and processor
clip_model = CLIPModel.from_pretrained("openai/
clip-vit-base-patch32")
clip_processor = CLIPProcessor.from_pretrained("openai/
clip-vit-base-patch32")
# Example product input
titles = ["Men's running shoes"]
images = [Image.open("shoe.jpg").convert("RGB")]
# Encode text and image
text_inputs = clip_processor(text=titles, return_
tensors="pt", padding=True, truncation=True)
image_inputs = clip_processor(images=images, return_
tensors="pt")
with torch.no_grad():
    text_emb = clip_model.get_text_features(**text_inputs)

image_emb = clip_model.get_image_features(**image_inputs)
# Normalize and average to get multimodal embedding
text_emb = text_emb / text_emb.norm(dim=-1, keepdim=True)
image_emb = image_emb / image_emb.norm(dim=-1, keepdim=True)
item_emb = (text_emb + image_emb) / 2
```

- **User Embeddings**: Computed via mean pooling of embeddings for items the user rated highly (≥ 4 stars).

Models Training and Evaluation

- **MLP Model**: A two-layer feedforward neural network that learns interaction from concatenated user-item embeddings as well as cosine similarity between user and item embeddings:

   Input (768+768) → Hidden (128) → Output (score)

- **DotProductModel**: Computes cosine similarity between user and item embeddings as the relevance score.

Table 6.3 Comparison between MLP and dot-product model in accuracy of pairwise ranking task

Model	Train accuracy	Test accuracy	Δ(Train-test)
MLP	80.5%	56.5%	24.0%
Dot product	82.5%	46.8%	35.7%

- **Loss Function**: BPR loss encourages the model to rank a positive item higher than a negative one:

```
# How BPR loss encourages proper ranking:
loss = -log(σ(pos_score - neg_score))   # σ=sigmoid
```

- Training Pairs: 1430 per user; 395 per user for testing.
- Evaluation Metric: We use pairwise accuracy, the percentage of correctly ranked item pairs as our evaluation metric.

6.4.3 Results and Analysis

The comparison results are shown in Table 6.3. While both models achieve similar training accuracy (~80%), their test performance diverges significantly:

- The *MLP model* achieves *56.5% test accuracy*, indicating it captures more generalizable patterns from the data.
- The *Dot-product model* overfits (train-test gap = 35.7%), performing poorly on unseen data. Its assumption of linear similarity fails to account for the subjective and multi-modal nature of fashion preferences.

These results emphasize that *simple embedding similarity* and even the more sophisticated *MLP model* is insufficient for nuanced domains like fashion, where taste depends on a combination of textual description, visual aesthetics, and user-specific signals.

6.5 Conclusion

This experiment highlights the limitations of dot-product models in subjective domains like fashion recommendation. Despite high training accuracy, the dot-product model generalizes poorly (46.8% test accuracy), indicating overfitting and an inability to capture non-linear user preferences. In contrast, a simple MLP achieves better generalization (56.5%), showing the value of even lightweight learned ranking functions when working with rich embeddings like CLIP.

While the accuracy remains modest, it reflects the complexity of modeling taste and style. These results point to the need for more expressive models that go beyond static similarity, paving the way for future work on user-conditioned ranking, hard negative sampling, and multi-modal fusion.

References

Alayrac, J.-B., Donahue, J., Luc, P., et al. (2022). *Flamingo: A visual language model for few-shot learning*. arXiv:2204.14198.

Arnab, A., Dehghani, M., Heigold, G., Sun, C., Lučić, M., & Schmid, C. (2021). *ViViT: A video vision transformer*. Proceedings of the IEEE/CVF International Conference on Computer Vision (ICCV) (pp. 6836–6846).

Baltrusaitis, T., Ahuja, C., & Morency, L.-P. (2019). Multimodal machine learning: A survey and taxonomy. *IEEE Transactions on Pattern Analysis and Machine Intelligence, 41*(2), 423–443.

Devlin, J., Chang, M.-W., Lee, K., & Toutanova, K. (2019). *BERT: Pre-training of deep bidirectional transformers for language understanding*. NAACL.

Gao, P., Ge, Y., Lin, J., Cai, J., Zhang, Y., & Zhang, Y. (2020). *FashionBERT: Text and image matching with adaptive loss for cross-modal retrieval*. Proceedings of the 43rd International ACM SIGIR Conference on Research and Development in Information Retrieval (SIGIR '20) (pp. 2251–2260).

He, K., Zhang, X., Ren, S., & Sun, J. (2016). *Deep residual learning for image recognition*. Proceedings of the IEEE Conference on Computer Vision and Pattern Recognition (CVPR) (pp. 770–778).

Hershey, S., Chaudhuri, S., Ellis, D. P. W., Gemmeke, J. F., Jansen, A., Moore, R. C., ... & Saurous, R. A. (2017). *CNN architectures for large-scale audio classification*. Proceedings of the IEEE International Conference on Acoustics, Speech and Signal Processing (ICASSP) (pp. 131–135).

Liu, J., Wang, Y., Zhou, C., Yu, J., Song, Y., & Zhang, Y. (2023). MMRec: Unified multimodal recommendation with generative multimodal transformer. *arXiv preprint arXiv:2305.13326*.

Lu, J., Batra, D., Parikh, D., & Lee, S. (2019). *ViLBERT: Pretraining task-agnostic visiolinguistic representations for vision-and-language tasks*. NeurIPS.

Radford, A., Kim, J. W., Hallacy, C., et al. (2021). *Learning transferable visual models from natural language supervision*. ICML.

Tsai, Y.-H. H., Bai, S., Yamada, M., et al. (2019). *Multimodal transformer for unaligned multimodal language sequences*. ACL.

Wei, T., Jin, B., Li, R., Zeng, H., Wang, Z., Sun, J., Yin, Q., Lu, H., Wang, S., He, J., & Tang, X. (2024). Towards unified multi-modal personalization: Large vision-language models for generative recommendation and beyond (arXiv:2403.10667v2). *arXiv*.

Chapter 7
Generative Recommendation and Planning Systems

This chapter examines the evolving landscape of *generative recommendation and planning systems*, which harness the capabilities of large language models (LLMs) to generate content, user profiles, and multi-step recommendation plans. We begin by exploring key content generation tasks such as personalized descriptions, summaries, and conversational responses. We then outline strategies for evaluating these outputs through benchmark construction and task-specific metrics. We then turn to *sequential planning*, where LLMs support multi-turn dialogue and goal decomposition to enable proactive, context-aware recommendation. Lastly, we introduce two practical tutorials: one on personalized profile generation, and another on multi-step task planning with recommendations.

7.1 Introduction

We have explored how *conversational recommendation systems* enable interactive preference elicitation through natural dialogue in Chap. 5, and how *multi-modal systems* enrich item and user understanding by integrating diverse data sources in Chap. 6. This chapter dives deep into the emerging class of *generative recommendation and planning systems (GRPS)*. Unlike traditional systems that retrieve or rank existing items, GRPS leverages large language models (LLMs) to *generate new content* in various formats, and *construct multi-step recommendation plans*.

7.1.1 Motivations

Traditional recommendation systems face three critical limitations:

1. **Static Content**: They recommend existing items but cannot generate novel content (e.g., a travel app can suggest hotels but cannot create a day-by-day itinerary).
2. **Limited Context Handling**: They struggle with multi-step, context-dependent tasks (e.g., balancing budget, time, and user preferences in a trip plan).
3. **Impersonalization**: They often fail to deliver truly unique experiences (e.g., generic product descriptions vs. AI-generated narratives tailored to user tastes).

Generative recommendation and planning systems address these gaps by:

- **Enabling Dynamic Content Creation**: For example, an e-commerce platform may generate personalized product descriptions like, "This rugged backpack is ideal for your hiking trips to the Rockies, with waterproof compartments for your camera gear."
- **Supporting Complex Decision-Making**: For example, a travel app may synthesize a 5-day itinerary for Paris, balancing cultural tours, dining, and leisure based on user preferences.
- **Delivering Hyper-Personalized Experiences**: For example, a music app may create a playlist *and* generate lyrics for a custom song reflecting the user's mood or memories.

These systems are particularly valuable in domains requiring *adaptability* (e.g., real-time travel adjustments), *creativity* (e.g., marketing campaigns), and *personalization* (e.g., coaching plans).

7.1.2 Content Generation Summary

Table 7.1 summarizes common generation tasks like text generation, image generation, audio and video generation, each with example use, popular models, and basic modeling principles.

Table 7.1 Summary of Models for Content Generation

Task	Example use	Popular models	Basic modeling principles
Text generation	Personalized product descriptions, emails	GPT-3/4, T5, BART	Auto-regressive (decoder-only), Encoder–Decoder
Image generation	Custom fashion or product visuals, ad creatives	DALL·E, Stable diffusion, Imagen	Diffusion models, Text-to-Image transformers, GANs
Audio generation	Personalized music generation, synthetic speech	WaveNet, Jukebox, AudioLM, Audio diffusion	Auto-regressive modeling, Spectrogram diffusion, Neural Vocoding
Video generation	Product demos, short-form content, trailers	Runway Gen-2, pika, Sora (by OpenAI)	Temporal diffusion, GANs, Text-to-Video transformers

7.1.3 *Text Generation*

Text generation plays a central role in personalized recommendation systems, enabling the dynamic creation of content such as product descriptions, summaries, reviews, and conversational responses. Modern generative models support a wide range of input types and control mechanisms, allowing outputs to be tailored to user intent, content structure, or stylistic requirements.

7.1.3.1 Categorization by Input Types

Text generation can be conditioned on a variety of input sources:

- **Natural Language Instructions**
  These are free-form prompts that describe the task to be performed, such as "Write a summary of this thread of emails" or "Explain how this product works in one sentence."
- **Structured Data Inputs**
  Structured fields such as product attributes, pricing, user profiles, or tabular data can be converted to text or used directly as input to the model. This is often implemented by serializing into JSON, CSV, or natural language form.
- **Dialogue Context or Interaction History**
  In conversational applications, models are conditioned on previous dialogue history or system actions to generate coherent and contextually appropriate responses.

7.1.3.2 Modeling Architectures

There are two dominant neural architectures that support text generation:

- **Decoder-Only Models**
  Auto-regressive models (e.g., GPT-2/3/4) generate tokens one by one, conditioning each new token on all previously generated ones. These models are especially effective for open-ended generation and in-context learning.
- **Encoder–Decoder Models**
  Models like T5 and BART first encode the input (e.g., instructions, metadata, or a document) into a latent representation, and then decode this into an output sequence. These models excel at input-conditioned generation tasks such as summarization or translation.

7.1.3.3 Controlled Generation Techniques

To steer model outputs toward specific formats, tones, or objectives, several control mechanisms can be applied:

- **Prompt Engineering**
  We can craft prompts to guide the model toward a desired behavior. This includes:
  - **Instructional Prompts**: "Write a short and friendly product description for a standing desk. Focus on comfortness, space efficiency, and the use of eco-friendly materials. The description should be easy to read and no longer than 4 sentences."
  - **Few-shot Examples**: Provide examples to guide style, tone, or structure (e.g., one product → one description format).
- **Template-Based Conditioning**
  Predefined templates or schema can be injected into the prompt to ensure output structure. For example, "Product: {name}. Category: {category}. Key Features: {features}."
- **Instruction Tuning**
  Fine-tuning the model on a large corpus of labeled instruction–response pairs (e.g., FLAN, Alpaca) helps improve reliability and control, especially for task-specific outputs.

7.1.4 Image Generation

Image generation systems enable machines to produce visual content from structured or unstructured inputs including text prompts, style references, etc. These models are increasingly used in design automation, product visualization, creative tools, and personalized media. Depending on the input modality and target use case, different architectures and control mechanisms can be employed to achieve high-quality and stylistical results.

7.1.4.1 Categorization by Input Types

- **Text-to-Image**
  The model generates an image from a descriptive prompt (e.g., "a wooden coffee table with a tall chair"). This is the most common use case in creative applications.

- **Image + Text**
  Used for *image editing, style transfer*, or *inpainting*, where an input image is modified based on an instruction (e.g., "make this outfit look more formal").
- **Latent Noise Vector**
  In GANs and diffusion models, generation starts from a random noise vector, which is iteratively transformed into a realistic image. This allows for diversity and controllability when paired with conditioning inputs.

7.1.4.2 Modeling Architectures

- **Diffusion Models**
  These models (e.g., *Stable Diffusion, Imagen*) gradually denoise a random input to generate high-fidelity images. They support fine-grained text conditioning and are widely used in open-source communities.
- **Generative Adversarial Networks (GANs)**
  GANs use a generator-discriminator setup to learn realistic image distributions (Goodfellow et al. 2014). While powerful, they are often harder to train and less interpretable than diffusion models.
- **Cross-Modal Transformers**
  Models like DALL·E (Ramesh et al., 2021) align text and image embeddings using transformer architectures, enabling strong semantic alignment between input prompts and output visuals. Datasets like LAION-5B (Schuhmann et al. 2022) have enabled the training of these models at web scale, making open-domain generation feasible across languages, domains, and styles.

7.1.4.3 Controlled Generation

- **Prompt Engineering (Text and Layout)**
  Prompts can specify content (a red leather armchair), composition (centered on a white background), or style (in Pixar animation style). Advanced systems even accept *layout sketches* or *bounding boxes* as additional control.
- **Reference-Based Control**
  By providing one or more reference images, the model can preserve style, structure, or identity. This is common in *avatar generation*, *concept art*, and *visual storytelling*.

Example: Text-to-Image with diffusers

Below is a code example using Hugging Face's diffusers library to generate an image from a simple prompt:

```
from diffusers import StableDiffusionPipeline
import torch
# Load the model (requires ~4GB VRAM, torch >= 1.13,
diffusers >= 0.11)
pipe = StableDiffusionPipeline.from_pretrained(
    "CompVis/stable-diffusion-v1-4",
    torch_dtype=torch.float16
)
pipe = pipe.to("cuda")  # Use GPU if available
# Define the prompt
prompt = "A futuristic headphone design, white and minimalistic"
# Generate the image
with torch.autocast("cuda"):
    image = pipe(prompt).images[0]
# Save the output
image.save("generated_headphones.png")
```

7.1.5 Audio Generation

Audio generation enables the creation of synthetic speech, sound effects, and even music. These can be created from various forms of input such as text, reference audio, or musical structure. Applications span from audiobook narration, conversational assistants to music generation and personalized voice agents. Unlike image generation, audio involves a fine-grained temporal component, which presents unique challenges for maintaining coherence, rhythm, and expressiveness over time. Advancements in TTS and neural audio modeling have significantly improved quality, control, and speaker adaptation.

7.1.5.1 Categorization by Input Types

- **Text-to-Speech (TTS)**
  Converts plain text into synthetic speech using neural vocoders. Common in voice assistants, audiobooks, and accessibility tools.
- **Voice Cloning**
  Takes a short reference audio clip and generates new speech in the same voice, useful for dubbing, localization, or custom avatars.

7.1 Introduction

- **Lyrics + Genre Tags**
  Used for singing voice generation and music synthesis (e.g., Jukebox, Dhariwal et al. 2020). Inputs may include lyrics, melody contours, and genre/style descriptions.
- **Audio-Text Alignment**
  Trained on aligned pairs of audio and transcripts (e.g., for speech synthesis, sound effect generation, or music-text modeling).

7.1.5.2 Modeling Architectures

- **Auto-regressive Waveform Models**
  Generate audio sample-by-sample (e.g., WaveNet by van den Oord et al. 2016), achieving high quality but with slow inference.
- **Diffusion-Based Audio Models**
  Generate audio via denoising in either the time or frequency domain. These models offer high fidelity and robustness.
- **Spectrogram + Vocoder Pipelines**
  A common architecture where the model first generates a mel-spectrogram (e.g., with Tacotron or Bark), which is then converted to waveform audio using a neural vocoder (e.g., HiFi-GAN, WaveGlow).

7.1.5.3 Controlled Generation

Modern TTS systems support various mechanisms to control the characteristics and expressiveness of generated speech. These controls typically fall into three categories:

Speaker Control

Speaker control focuses on *who* is speaking, and it controls the speaker's voice characteristics using *speaker embeddings, reference recordings,* or *ID tokens*. Speaker control enables *voice cloning, multi-voice synthesis,* or *persona creation* in multilingual and conversational systems.

Prosody and Emotion Control

Prosody and emotion control focuses on *how* something is spoken.

- Prosody includes *rhythm, pitch, speed*, and *intonation*—key elements for expressive and natural-sounding audio.
- Controlled using:
  - *Latent variables* for prosody and emotional style

- *Acoustic feature conditioning* (e.g., pitch contours, energy levels)
- *Token-level markup* such as SSML (Speech Synthesis Markup Language) for rule-based prosody control\

SSML Example

```
<speak>
  Hello! <prosody rate="slow" pitch="+10%">I'm here to help you find the perfect gift.</prosody>
</speak>
```

- For fine-grained control, models like EmoCtrl-TTS (Zhang et al., 2023) allow *phoneme-level emotional conditioning*, enabling dynamic emotional variation across an utterance.

Content and Prompt-Level Control

Content and prompt-level control focuses on *What* nonverbal or stylistic content is included.

- Systems like Bark can interpret rich prompts that mix text with *sound effects, emojis,* or *musical symbols* to enrich the expressive output.
- For example, including [laughter], or [clears throat] in the prompt leads to corresponding audio events.

7.1.5.4 Example: TTS with Bark (Suno AI)

This example uses the open-source bark library to generate expressive speech from text.

```
from bark import SAMPLE_RATE, generate_audio
import scipy.io.wavfile as wavfile
# Define the prompt
prompt = "Welcome to your personalized shopping assistant. Let's find something great for you today!"
  #
  Generate
  audio with the default speaker ("v2/en_speaker_6")
audio_array = generate_audio(prompt)
# Save to a WAV file
wavfile.write("speech.wav", SAMPLE_RATE, audio_array)
```

7.1.6 Video Generation

Video generation brings together spatial and temporal modeling to create coherent, visually rich sequences. We can generate videos from a wide range of inputs, including text, images, or structured scene plans. Video synthesis enables applications in marketing, storytelling, education, and social media. Compared to image generation, video models must learn to capture motion, scene transitions, and character consistency across frames, making them both computation-intensive and architecturally more complex.

7.1.6.1 Categorization of Input Types

- **Text-to-Video**
  Generate a complete video from a natural language description (e.g., "a person surfing at sunset"). Useful for storyboard prototyping and creative scene generation.
- **Image-to-Video**
  Animate a static image or portrait using motion cues (e.g., facial landmarks, pose trajectories), often used in avatar animation and talking head generation.
- **Storyboard-to-Video**
  Use structured multi-scene input (e.g., keyframes, scene descriptions, and temporal order) to guide transitions and video composition.
- **Motion or Pose Input**
  Provide body or object motion data (e.g., OpenPose keypoints) to animate characters or simulate realistic movement.

7.1.6.2 Modeling Architectures

- **Temporal GANs**
  Models like MoCoGAN (Tulyakov et al., 2018) separate motion and content streams to generate videos frame-by-frame, enabling realistic temporal dynamics. Often used for short, stylized clips.
- **Hierarchical VAEs or VQ-Based Models**
  Compress spatial and temporal components using discrete representations (e.g., VQ-VAE-2 by Razavi et al. 2019, TATS), supporting scalable video generation.
- **Transformer-Based Video Models**
  Use spatiotemporal attention mechanisms to model long-range dependencies. Models like Make-A-Video (Singer et al., 2022) extend diffusion and transformer architectures to the video domain, generating coherent temporal sequences

directly from text without requiring paired text-video datasets. These models offer better temporal coherence and prompt alignment.

7.1.6.3 Controlled Generation Techniques

- **Prompt-Based Content Control**
  Text prompts specify high-level attributes such as characters, setting, objects, and mood.
- **Motion Trajectory or Flow Control**
  Define how objects or people move across frames, or simulate specific camera motions.
- **Temporal Conditioning**
  Adjust the duration, frame pacing, or scene transitions to control storytelling rhythm or visual tempo.

7.1.6.4 Script-to-Video Example: Intelligent Museum Narrative

The diagram in Fig. 7.1 illustrates a modular pipeline for generating educational or storytelling videos from user-provided input such as text, images, or script. The system first plans the scenes, then generates corresponding narration and visuals using text-to-speech (XTTS) and image generation models (e.g., DALL·E or Stable Diffusion). All components are merged in a final assembly step to produce exportable videos or editable presentation slides.

Suppose we aim to generate a short animated video set in a modern museum. The narrative centers around a sentient statue that awakens and encounters an AI-powered assistant. Through their interaction, the assistant explains how it enhances the museum experience by offering voice-guided and visual explanations for each exhibit. Together, they envision a future where museums become interactive, intelligent environments tailored to each visitor.

The story is defined through a structured scene-based script represented as a JSON object, where each scene contains a high-level description, corresponding narration, speaker identifier, and an image generation prompt. The structure aligns with the pipeline shown in Fig. 7.1, where narration and visual elements are generated independently, then assembled into final video output.

7.1 Introduction

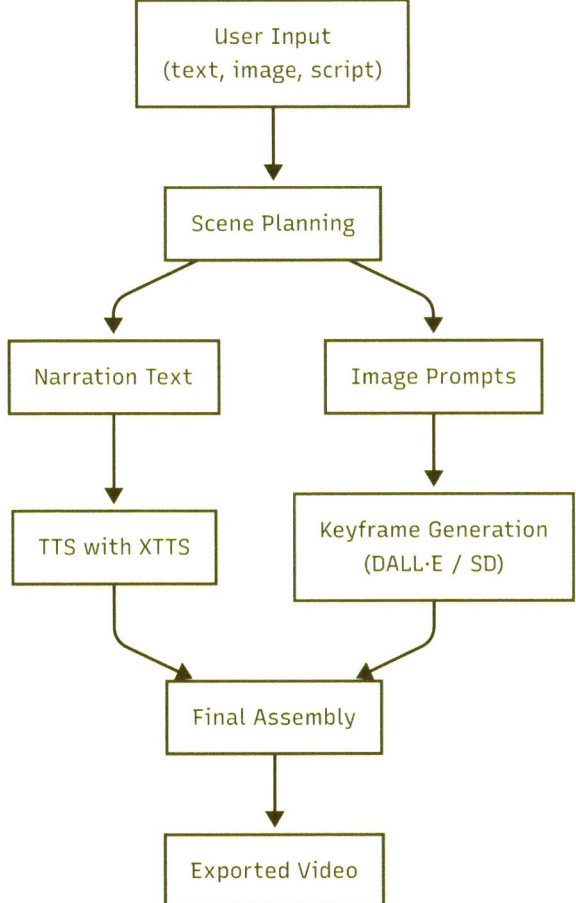

Fig. 7.1 Script-to-Video Generation Pipeline

Scene Script (JSON Format)

```
[
  {
    "scene_id": 1,
    "scene_description": "Inside a quiet, dimly lit museum hall, a spotlight shines on an ancient statue surrounded by artifacts.",
    "narration": "Where am I... and why is everything so quiet?",
    "speaker_id": "statue",
    "image_prompt": "an ancient museum hall at night,
```

```
spotlight on a stone statue surrounded by glass display
cases, cinematic lighting"
  },
  {
    "scene_id": 2,
    "scene_description": "The statue slowly comes to life,
blinking and stepping off its pedestal.",
    "narration": "I feel... awake. Have I been asleep for
centuries?",
    "speaker_id": "statue",
    "image_prompt": "a stone statue animated to life,
stepping off a pedestal in a museum, dramatic shadows and
subtle animation"
  },
  {
    "scene_id": 3,
    "scene_description": "A futuristic assistant appears,
hovering with a soft glow, greeting the statue.",
    "narration": "Hello! I'm your virtual assistant, here to
help visitors explore and learn.",
    "speaker_id": "assistant",
    "image_prompt": "a glowing AI assistant hovering near a
statue, modern and friendly design, set inside a museum"
  },
  {
    "scene_id": 4,
    "scene_description": "The assistant explains how it
helps guide visitors through voice and interactive
visuals.",
    "narration": "I answer questions, share stories, and
adapt to every guest's curiosity.",
    "speaker_id": "assistant",
    "image_prompt": "a digital interface projected from the
assistant, displaying museum info and guiding visuals,
futuristic UI"
  },
  {
    "scene_id": 5,
    "scene_description": "The statue listens intently,
intrigued by the assistant's capabilities.",
    "narration": "Impressive. I never imagined exhibits
could talk back.",
    "speaker_id": "statue",
```

7.2 Evaluation

```
    "image_prompt": "a stone statue looking curious and
thoughtful, facing a floating assistant in a high-tech museum
gallery"
  },
  {
    "scene_id": 6,
    "scene_description": "They both look around the hall as
screens illuminate and visitors appear.",
    "narration": "Together, we'll bring the past to life for
every visitor.",
    "speaker_id": "assistant",
    "image_prompt": "museum hall lighting up with
interactive displays and visitors arriving, the assistant
and statue in the foreground"
  }
]
```

Using this structured script, we proceed through the pipeline in Fig. 7.1:

1. Scene Planning decomposes the story into individual segments.
2. Narration Text is converted into natural speech using multilingual TTS (e.g., XTTS).
3. Image Prompts guide keyframe generation via tools like DALL·E or Stable Diffusion.
4. All assets are then *assembled* into short video segments using image-to-video models.
5. Finally, segments are stitched together and exported as a coherent short video or editable presentation.

7.2 Evaluation

Evaluating the quality of generative content requires a structured approach to *benchmark design, metric selection*, and *evaluation methods*. This section extends prior discussions on evaluation. Section 2.1.5 covered core recommendation metrics such as relevance, diversity, and novelty, while Sect. 3.4 introduced LLM-based evaluation methods, including LLM-as-a-judge and synthetic data generation. Section 5.2.4 focused on conversational systems, highlighting continuous evaluation of coherence, user satisfaction, and interaction success.

Generative content evaluation shares the same fundamental principle, but introduces another set of challenges:

- **Relevance and Personalization**: Shared with retrieval and recommendation tasks, but harder to assess when outputs are diverse or subjective.
- **Fluency and Coherence**: Especially important in generated text and dialogue, echoing metrics from conversational recommendation.
- **Cross-Modal Alignment**: For image, audio, and video generation, evaluation must capture how well outputs match the conditioning prompt or input modality.

7.2.1 Constructing Benchmark Data

Benchmark datasets are essential for evaluating generative systems, as they provide standardized tasks and consistent inputs for comparison. A crucial first step is to *define the evaluation task*, which includes specifying the *input-output structure* and the *evaluation target*.

For example,

- In *text summarization*, the evaluation task consists of a source article (input) and a reference summary (target), against which generated summaries are compared.
- In *text-to-image generation*, the input is a descriptive prompt, and the output is evaluated against reference images or through human judgments.

After task definition, the next step is *data collection*, which involves gathering inputs that reflect realistic application scenarios. We can use a combination of publicly available data, AI-generated data and data from applications to form our evaluation set:

- *Public datasets* like MS-COCO for text-image embedding alignment or image retrieval (Lin et al. 2015) or LibriSpeech for speech recognition (Panayotov et al. 2015).
- *Synthetic data* created using LLMs or generative models to easily scale up the evaluation dataset.
- *User-generated content* from real applications represents real-world distributions (e.g., actual queries or voice commands).

7.2.2 Dimensions and Metrics

Evaluating generative models requires a multifaceted approach. Metrics vary across *modalities* (text, image, audio, video) and *evaluation goals* such as fidelity, relevance, diversity, and safety. Table 7.2 lists key dimensions and metrics for each generation task.

7.2 Evaluation

Table 7.2 Key dimensions and metrics for each generation task

Task	Dimension	Metrics
Text generation	Relevance	BLEU, ROUGE, BERTScore
	Coherence	Perplexity
	Ethical considerations	Toxicity detection, Fairness metrics
Image generation	Fidelity	FID, IS
	Relevance	CLIPScore
	Diversity	LPIPS
Audio generation	Quality	PESQ, STOI
	Relevance	Embedding similarity
Video generation	Temporal consistency	FVD
	Relevance	CLIPScore

We group metrics by their core evaluation objective, and highlight representative metrics like classical BLEU and Perplexity.

7.2.2.1 Fidelity and Quality

These metrics assess how realistic, coherent, or high quality the generated outputs are compared to reference data.

- **Text**:
  - *Perplexity* (Bengio et al., 2003): Measures how well a language model predicts word sequences.

    $$\text{Perplexity} = \exp\left(-\frac{1}{N} \sum_{i=1}^N \log P(w_i \mid w_1, \dots, w_{i-1})\right)$$

    Lower perplexity suggests more fluent and coherent outputs, while higher perplexity suggests uncertainty or poor performance.

- **Image**:
  - FID (Fréchet Inception Distance): Compares feature statistics between real and generated images.

    $$\text{FID} = |\mu_r - \mu_g|^2 + \text{Tr}(\Sigma_r + \Sigma_g - 2(\Sigma_r \Sigma_g)^{1/2})$$

  - **Audio**:
    - PESQ (Perceptual Evaluation of Speech Quality): Compares reference and synthesized audio using perceptual models.
    - FVD (Fréchet Video Distance): Extension of FID for video, accounting for temporal coherence across frames.

7.2.2.2 Relevance and Alignment

These metrics evaluate whether the generated output aligns with the input prompt, context, or user intent.

- **Text**:
  - BLEU (Papineni et al., 2002): Measures n-gram overlap with reference text.
    $$\text{BLEU} = BP \cdot \exp\left(\sum_{n=1}^N w_n \log p_n\right)$$
    Where BP is the brevity penalty and p_n are n-gram precisions.
  - BERTScore (Zhang et al., 2020): Uses BERT embeddings to assess semantic similarity between generated and reference text. Unlike BLEU, which relies on n-gram overlap, BERTScore captures semantic meaning similarity even when the wording differs significantly.

- **Multi-modal (Text ↔ Image/Video)**:
  - CLIPScore: Uses CLIP embeddings to measure alignment between generated media and textual prompts.

7.2.2.3 Diversity and Expressiveness

These metrics test whether the model can produce varied, rich outputs across different prompts or over multiple generations.

- **Text**:
  - **Distinct-n**: Measures the proportion of unique n-grams in generated text.
  - **Self-BLEU**: Computes BLEU between multiple generated outputs to detect redundancy.
- **Image**:
  - **LPIPS (Learned Perceptual Image Patch Similarity)**: Assesses perceptual dissimilarity between pairs of generated images to quantify diversity, using deep network features to approximate human visual similarity judgments.

7.2.2.4 Safety and Toxicity

Generative systems must avoid producing harmful, biased, or offensive outputs, especially in public-facing deployments.

- **Toxicity Detection**:
  - Perspective API or similar tools classify harmful language (e.g., hate speech, profanity) in text outputs.
- **Fairness Metrics**:
  - Evaluate whether the model treats demographic groups equitably.
  - Example: Measuring output sentiment or exposure balance across gender or race categories in recommendations or generation.

7.2.3 Evaluation Method

Evaluating generative content involves both automated metrics and human judgment:

1. **LLM-as-a-Judge**:
   - Use LLMs to evaluate the quality of generated content by comparing it to ground truth or predefined criteria.
   - *Example*: GPT-4 can assess the coherence and relevance of generated text or provide feedback on image descriptions.
   - **Advantages**: Scalable, cost-effective, and consistent.
   - **Limitations**: May lack nuanced understanding or contextual awareness.

2. **Human Expert Judgment**:
   - Employ domain experts to evaluate content quality based on subjective criteria (e.g., creativity, aesthetic appeal).
   - *Example*: Artists rate the visual quality of generated images, or writers assess the narrative flow of generated text.
   - **Advantages**: Captures nuanced, context-aware evaluations.
   - **Limitations**: Time-consuming, expensive, and prone to subjectivity.

3. **Hybrid Approaches**:
   - Combine LLM-as-a-judge with human evaluation to balance scalability and depth.
   - *Example*: Use LLMs for initial screening and humans for final validation.

7.3 Sequential Planning with LLMs

Traditional recommendation systems focus on predicting the next-best item, but generative sequential planning introduces a paradigm shift: using LLMs to generate coherent, multi-step plans that align with user preferences, context, and constraints.

Importantly, this paradigm also emphasizes *verifiable outcomes*, enabling systems to not only propose plans but also justify and evaluate them using external signals or tools.

7.3.1 Key Components

7.3.1.1 Sequential Decision-Making

LLMs excel at modeling multi-step decision sequences by leveraging long-range dependencies and contextual understanding. Rather than selecting a single next-best item, they can reason over extended trajectories. For example: "watch a movie → dine at a restaurant → attend a concert" to create cohesive and context-aware experiences. Unlike traditional Markovian or shallow models that assume limited memory or independence between steps, LLMs can incorporate rich signals such as prior preferences, temporal patterns, spatial constraints, and latent goals.

7.3.1.2 Planning as Constrained Generation

LLMs can treat recommendation planning as a *constrained generation task,* where the output must satisfy a set of user-defined or system-imposed conditions. These constraints can include:

- **Hard constraints**: Budget caps, location bounds, time windows
- **Soft constraints**: Genre preferences, novelty goals, diversity targets
  For instance, generating a travel itinerary that maximizes adventure-related activities while staying under a $500 budget requires the model to reason about item compatibility, cost aggregation, and user preferences in tandem.

7.3.1.3 Dynamic Adaptation

Real-world preferences are dynamic: users change their minds, revise their preferences, or refine their goals mid-way. LLMs can incorporate *interactive feedback* and adjust previously generated plans accordingly. This adaptability is key for practical deployment:

For example: After suggesting a museum tour, a user might say, "I'm tired of indoor activities—can you recommend something outdoors?" The model can revise the plan dynamically, replacing or reordering steps while maintaining consistency.

7.3 Sequential Planning with LLMs

This capability stems from combining generative planning (structured text output) with retrieval augmentation (e.g., recommending specific items from a catalog), allowing LLMs to act as *both planners and adapters*.

7.3.1.4 Verifiable Outcomes

LLM-generated plans must be *valid, feasible, and aligned* with real-world constraints. To ensure this, external tools or functions can *verify outcomes* post-generation:

- Is the total cost within the allowed budget?
- Does the plan exceed time constraints?
- Are selected items available or compatible?

Verification can be performed via *external simulation, rule-based checks*, or *differentiable constraints*, which provide transparency and trustworthiness. This layer also enables model debugging and post-hoc editing for high-stakes scenarios like healthcare planning, curriculum design, or financial advising.

7.3.2 Application Scenarios

A compelling use case for multi-step planning is *personalized project planning with verifiable constraints.*

Scenario: A user wants to build a home gym but is unsure how to allocate space, select equipment, and stay within budget.

Plan Generation: The LLM generates a coherent, step-by-step setup:

1. Assess available space (e.g., 100 sq ft).
2. Recommend compact, multipurpose equipment.
3. Suggest layout configurations.
4. Provide a purchasing plan under $2000.

Outcome Verification: External functions evaluate the feasibility:

- Total cost check: Is it under budget?
- Space simulation: Do selected items fit?
- Preference alignment: Does the plan match fitness goals?

This hybrid planning-verification approach demonstrates how LLMs can support *goal-driven, constraint-aware recommendations* across domains—ranging from home projects to career planning and event logistics.

7.3.2.1 Example of External Verification

```
def verify_budget(plan, budget):
    # Extract total cost from the plan (simulated)
    total_cost = 1650   # Example value extracted from the plan
    if total_cost <= budget:
        return "Budget verification passed: Plan is within budget."
    else:
        return "Budget verification failed: Plan exceeds budget."
budget = 2000
verification_result = verify_budget(plan, budget)
print(verification_result)
```

This scenario highlights the practical utility of LLMs in generating *structured, interpretable,* and *verifiable* multi-step plans—making them ideal for real-world applications like home improvement, travel planning, or educational curriculum design.

7.4 Tutorial: Image-to-Avatar Generation

7.4.1 Overview

In this tutorial, we explore how to generate personalized avatars from real face images using generative models, specifically leveraging image-to-image diffusion pipelines like Stable Diffusion. This task provides an intuitive and visual entry point into the world of *multi-modal generation*, where inputs span different modalities (images, text prompts) and outputs are highly stylized image content.

Goal of this Tutorial

- Preprocess and condition images for generation tasks
- Apply Stable Diffusion's img2img pipeline for stylistic avatar generation
- Evaluate generation quality using both *identity preservation* and *style matching* metrics

We show a condensed version of this tutorial in the book text. The full version of the code is available at: https://github.com/qqwjq1981/springer-LLM-recommendation-system

7.4.2 Experimental Design

7.4.2.1 Data Source

We use a subset of the Flickr-Faces-HQ (FFHQ) dataset (Karras et al. 2019), a high-quality image collection of aligned human faces spanning age, ethnicity, and facial features. This dataset is publicly available via Hugging Face and licensed for research use.

- Dataset: FFHQ (nateraw/ffhq)
- Size: 100 images (for tractability)
- Preprocessing: Resize to 512 × 512 resolution

7.4.2.2 Methods

We apply a Stable Diffusion-based img2img pipeline:

- **Base model**: runwayml/stable-diffusion-v1-5

```
from diffusers import StableDiffusionImg2ImgPipeline
pipe    =    StableDiffusionImg2ImgPipeline.from_
pretrained("runwayml/stable-diffusion-v1-5").
to("cpu")
init_image    =    Image.open(path).convert("RGB").
resize((512, 512))
result  =  pipe(prompt=prompt,  image=init_image,
strength=0.75, guidance_scale=7.5).images[0]
```

- **Prompt control**: We consider two prompt versions: Pixar-style (expressive, playful) Ghibli-style (2D, gentle, anime-inspired)

```
def get_prompt_from_version(prompt_version):
    if prompt_version == "Ghibli":
        return "A Studio Ghibli-style portrait that closely
resembles the original person, soft lighting, gentle colors,
2D anime-style illustration"
    else:
        return "Pixar-style character portrait, clean
features, cute and friendly expression, high quality
digital art"
```

7.4.2.3 Evaluation Metrics

As illustrated in Table 7.3, we evaluate how well these generated avatars:

- Maintain visual fidelity (FID score)
- Preserve identity (face embedding similarity)
- Align with the prompt (CLIP score)

Dark or empty images were automatically excluded from evaluation.

7.4.3 Results and Analysis

Figure 7.2 showcases five example subjects across three rows: original human portraits (top), Pixar-style avatars (middle), and Ghibli-style avatars (bottom). Pixar-style outputs preserve identity better and exhibit higher visual alignment with prompts, while Ghibli-style avatars introduce more abstraction and artistic variance, often deviating from original facial features.

Table 7.4 displays evaluation metrics on 100 style-transferred images, after excluding failed transfer. Here are some key takeaways from the metrics-based evaluation:

Table 7.3 Evaluation metrics for image-to-avatar generation

Metric	Description
FID (↓)	Frechet Inception Distance—measures distributional distance to original images
Identity (↑)	Cosine similarity between face embeddings of raw vs. generated image
CLIP score (↑)	Cosine similarity between prompt and image embeddings using CLIP

Fig. 7.2 Comparison of Raw Portraits and Stylized Avatars in Pixar and Ghibli Styles

Table 7.4 Evaluation metrics for image-to-avatar generation

Style	FID ↓	Identity ↑	CLIP Score ↑	Excluded Dark Images
Pixar	**191.64**	**0.085**	**0.316**	8
Ghibli	246.55	0.043	0.289	5

1. Pixar-style avatars outperformed Ghibli-style across all metrics:
   - Lower FID (closer to natural image statistics).
   - Higher identity preservation.
   - Better prompt-image alignment.
2. Ghibli-style avatars exhibit artistic abstraction, but this comes at a cost of losing facial resemblance.
3. A small percentage of generated avatars were completely dark or failed, filtering these improves metric robustness. In production, we should use re-try to make the pipeline more robust.

7.4.4 Discussion

This tutorial demonstrated stylized avatar generation using diffusion models in Pixar and Ghibli styles, evaluated via FID, CLIPScore, and *identity similarity*.

Pixar-style avatars outperformed Ghibli in FID (191.6 vs. 246.5) and identity preservation (0.085 vs. 0.044), suggesting stronger facial consistency and prompt alignment. Some dark or invalid outputs were filtered (8 Pixar, 5 Ghibli), indicating a need for robustness checks.

Next Steps

- *Model fine-tuning* to improve identity preservation beyond prompt engineering
- *Post-filtering* to remove artifacts or low-quality generations
- *Controllable stylization* or *style fusion* to allow user-specific customization

This lays the foundation for building user-personalized, style-aware avatar generation systems.

7.5 Second Tutorial: Goal-Driven Planning with LLMs

7.5.1 Overview

This tutorial demonstrates how to decompose a complex user goal, such as building a home gym, into a structured, actionable plan using an LLM. The system takes user-specific constraints as input, generates purchase plans, and uses both programmatic and LLM-based methods for verification and evaluation.

Goal of this Tutorial

- Learn how to break down open-ended goals into multi-step plans using LLMs.
- Generate structured recommendations under real-world constraints.
- Evaluate the plan's quality using both tool-based checks and LLM-as-a-judge feedback.

Key Features

- **Context Integration**: Personalizes output using inputs like budget, space, fitness level, and preferences.
- **Structured Output**: Returns plans in JSON format for easy downstream usage or validation.
- **LLM-Based Evaluation**: Assesses coherence, relevance, and personalization using an LLM critic.

We show a condensed version of this tutorial in the book text. The full version of the code is available at: https://github.com/qqwjq1981/springer-LLM-recommendation-system

7.5.2 Experimental Design

7.5.2.1 User Constraint Specification

We define a user scenario with basic constraints:

```
{
  "goal": "build a home gym",
  "budget": 2000,
  "space": "10 ft x 12 ft",
  "fitness_level": "intermediate",
  "preferences": ["cardio", "compact equipment"]
}
```

Step-by-Step Design

1. **Plan Generation**:
   GPT-4o is prompted with user constraint to generate a structured home gym plan in JSON, including layout and equipment suggestions.

7.5 Second Tutorial: Goal-Driven Planning with LLMs

```
planner_prompt = ChatPromptTemplate.from_messages([
    ("system", "You are a planning assistant that returns
JSON plans for home gym setup."),
    ("human", """Given the user's profile:
- Budget: ${budget}
- Room Size: {space}
- Fitness Level: {fitness_level}
- Preferences: {preferences}
Generate a JSON plan with:
- space_plan: string
- equipment: list of {{ "name": ..., "price": ... }}
- setup_notes: string""")
])
```

2. **Verification (Simulated Tool Use)**:
   Simple Python functions are used to verify that the plan:
   - Stays within budget
   - Uses space-efficient equipment as described

3. **Evaluation**:
   A separate LLM agent reviews the plan and provides qualitative feedback based on coherence, relevance, and personalization.

```
critic_prompt = ChatPromptTemplate.from_messages([
    ("system", "You are an evaluator that gives 1-10 scores
for coherence, relevance, and personalization."),
    ("human", "Evaluate the following plan:\n\n{plan_json}")
])
```

7.5.3 Results and Analysis

7.5.3.1 Generated Plan (Excerpt)

```
{
  "space_plan": "Arrange the equipment along the longer wall
to maximize space. Leave a 3 ft wide path for movement and
stretching. Use vertical storage solutions for smaller
items.",
  "equipment": [
    { "name": "Folding Treadmill", "price": 600 },
    { "name": "Compact Rowing Machine", "price": 500 },
    { "name": "Adjustable Dumbbells Set", "price": 300 },
    { "name": "Resistance Bands Set", "price": 50 },
    { "name": "Wall-Mounted Pull-Up Bar", "price": 100 },
    { "name": "Yoga Mat", "price": 30 },
    { "name": "Compact Exercise Bike", "price": 400 }
  ],
  "setup_notes": "Focus on compact and foldable equipment to
save space... [truncated]"
}
```

7.5.3.2 Tool-Based Verification

- **Budget**: Total = $1980 < $2000
- **Space**: Majority of items are compact or foldable

7.5.3.3 LLM-as-a-Judge Evaluation

Table 7.5 summarizes LLM evaluations of the generated home gym plan across three key metrics. Each score is supported by a justification, providing insight into how well the plan meets expectations for structure, relevance, and personalization. While

Table 7.5 LLM-judged metric scores with justification

Metric	Score	Justification
Coherence	9/10	Logical flow from layout to equipment; setup notes support spatial reasoning
Relevance	9/10	Recommendations match cardio and strength training within defined limits
Personalization	7/10	Addresses constraints well, but lacks tailored advice for fitness level

coherence and relevance scored highly (9/10), personalization showed room for improvement.

7.5.4 Discussion

This tutorial demonstrates how LLMs can generate structured, actionable plans using chain-of-thought reasoning. By emitting JSON-formatted outputs, the system supports programmatic validation and downstream consumption. Constraint-checking tools verify objective feasibility, while LLM-as-a-judge scoring adds nuanced subjective evaluation.

Strengths

- High-quality, interpretable output via prompt engineering
- Compatible with automated validation and refinement pipelines
- Supports iterative enhancement through multi-agent workflows

Limitations

- LLM-based critics may hallucinate or miss feasibility gaps
- Multi-agent orchestration (e.g., via LangChain) introduces complexity

Takeaway

- Combining generation, verification, and critique yields robust, modular recommendation pipelines. Even minimal tool integration boosts reliability and user trust when evaluation is systematic and structured.

References

Bengio, Y., Ducharme, R., Vincent, P., & Jauvin, C. (2003). A neural probabilistic language model. *Journal of Machine Learning Research, 3*, 1137–1155.

Dhariwal, P., Jun, H., Payne, C., Kim, J. W., Radford, A., & Sutskever, I. (2020). *Jukebox: A generative model for music* (arXiv:2005.00341). arXiv.

Goodfellow, I., Pouget-Abadie, J., Mirza, M., et al. (2014). *Generative adversarial nets*. NeurIPS.

Karras, T., Laine, S., & Aila, T. (2019). *A style-based generator architecture for generative adversarial networks*. CVPR.

Lin, T.-Y., Maire, M., Belongie, S., Bourdev, L., Girshick, R., Hays, J., Perona, P., Ramanan, D., Zitnick, C. L., & Dollár, P. (2015). *Microsoft COCO: Common objects in context* (arXiv:1405.0312v3). arXiv.

Panayotov, V., Chen, G., Povey, D., & Khudanpur, S. (2015). *Librispeech: An ASR corpus based on public domain audio books*. Proceedings of the 2015 IEEE International Conference on Acoustics, Speech and Signal Processing (ICASSP) (pp. 5206–5210). IEEE.

Papineni, K., Roukos, S., Ward, T., & Zhu, W. J. (2002). *BLEU: A method for automatic evaluation of machine translation*. Proceedings of the 40th Annual Meeting of the Association for Computational Linguistics (pp. 311–318).

Ramesh, A., Pavlov, M., Goh, G., Gray, S., Voss, C., Radford, A., Chen, M. Sutskever, I. (2021). *Zero-shot text-to-image generation.* Proceedings of the 38th International Conference on Machine Learning, in Proceedings of Machine Learning Research (vol. 139, pp. 8821–8831).

Razavi, A., van den Oord, A., & Vinyals, O. (2019). Generating diverse high-fidelity images with VQ-VAE-2. .

Schuhmann, C., Beaumont, R., Vencu, R., Gordon, C., Schilling, J., Kaczmarczyk, R., Clark, A., Serban, A., & Brüggemann, D. (2022). *LAION-5B: An open large-scale dataset for training next generation image-text models.* arXiv preprint arXiv:2210.08402.

Singer, U., Polyak, A., Shechtman, E., et al. (2022). *Make-a-video: Text-to-video generation without text-video data.* arXiv:2209.14792.

Tulyakov, S., Liu, M.-Y., Yang, X., & Kautz, J. (2018). *MoCoGAN: Decomposing motion and content for video generation.* CVPR.

van den Oord, A., Dieleman, S., Zen, H., et al. (2016). *WaveNet: A generative model for raw audio.* arXiv:1609.03499.

Zhang, T., Kishore, V., Wu, F., Weinberger, K. Q., & Artzi, Y. (2020). *BERTScore: Evaluating text generation with BERT.* Proceedings of the 8th International Conference on Learning Representations (ICLR).

Zhang, Z., Liu, W., Wang, Y., Liang, D., Wu, Y., & Liu, Y. (2023). *EmoCtrl-TTS: Fine-grained emotional control for text-to-speech synthesis using emotion factor modeling.* Proceedings of the 61st Annual Meeting of the Association for Computational Linguistics (ACL 2023).

Chapter 8
Challenges and Trends in LLMs for Recommendation Systems

This chapter offers a forward-looking perspective on the evolution of recommendation systems, highlighting emerging trends and open challenges. We focus on five key research frontiers: multi-modal integration, verifiable outcomes, multi-agent systems, generative copyright and privacy, and ethical AI and fairness. For each frontier, we illustrate not only the challenges it presents but also promising directions for advancing next-generation LLM-powered recommenders.

8.1 Introduction

As large language models (LLMs) increasingly power modern recommendation systems, new opportunities and challenges are emerging. This chapter explores *five key research frontiers* that we believe will shape the next generation of LLM-driven recommenders, reflecting both practical demands and open questions in the field.

- **Multi-modal Integration**: With content and interactions spanning text, images, audio, and video, integrating multiple modalities is crucial for capturing user preferences and context more accurately.
- **Verifiable Outcomes**: As generative recommenders move beyond static item lists toward dynamically generated plans, narratives, or multi-step suggestions, verifying the quality, relevance, and trustworthiness of outputs becomes a central challenge. This includes validating whether generated recommendations satisfy user constraints, align with stated goals, and are supported by verifiable evidence or reasoning.
- **Multi-agent Systems**: Coordinating multiple LLM-based agents for retrieval, planning, or dialogue offers a path toward more dynamic, goal-oriented recommendation experiences.

- **Generative Copyright and Privacy**: As systems begin generating content, new concerns arise around content ownership, user data protection, and responsible use of synthetic content.
- **Ethical AI and Fairness**: Ensuring transparency, fairness, and inclusivity remains essential for building trusted and socially responsible recommendation systems.

These frontiers highlight the complexity and promise of the future: systems that are not only intelligent and personalized, but also explainable, secure, and ethically aligned.

8.2 Multi-modal Integration

8.2.1 Challenges

8.2.1.1 Data Alignment

Data alignment refers to the challenge of bringing heterogeneous data types, such as text, images, audio, and video, into a shared semantic space where they can be jointly understood and compared. Each modality has its own structure and encoding: text is symbolic and sequential, images are spatial, and audio is temporal and continuous.

Aligning these disparate forms into meaningful, unified embeddings is non-trivial, particularly when the information is incomplete or only weakly correlated across modalities. Effective alignment requires cross-modal representation learning techniques that preserve the semantics of each modality while allowing for comparison and fusion (Baltrušaitis et al., 2019).

For example, in a fashion recommendation system, the model must understand that a review stating "sleek red boots" corresponds to a product image showing bright red ankle boots, even if the word "ankle" wasn't used in the text.

8.2.1.2 Consistency Across Modalities

While data alignment ensures that different modalities are meaningfully matched, consistency across modalities is about ensuring that *their contributions to a recommendation agree* or reinforce one another rather than contradict. Multi-modal systems may receive conflicting signals. For example, a movie that is described as "heartwarming" in reviews but features a poster with dark, eerie visuals. In such cases, the system must decide how to interpret and reconcile these differences, rather than treating all modalities equally or independently.

For example, a user interested in light-hearted romantic comedies may encounter a recommendation for a film whose *text description appears upbeat*, while its *trailer conveys a somber or violent tone*. This kind of *modal inconsistency* can lead to user confusion or dissatisfaction. A *consistency-aware multi-modal system* would detect this discrepancy between modalities, weigh it against the user's intent, and either *seek clarification* or *deprioritize the item* in ranking to ensure trustworthy recommendations.

8.2.1.3 Computational Complexity

Besides data alignment and consistency across modalities, *computational complexity* presents major challenges for leveraging multi-modal data in recommendation systems, especially in real-time scenarios. Processing spatially rich inputs such as high-resolution images or video frames demands significant GPU resources. Tasks like visual feature extraction, temporal modeling, and cross-modal fusion are computationally intensive and can introduce latency incompatible with low-latency requirements. Real-time adaptation to user behavior may further require reprocessing or reranking, compounding the load.

Liang et al. (2023) underscore these challenges in high-modality systems, showing that as more diverse modalities are added, the computational burden scales rapidly. They propose metrics to quantify modality and interaction heterogeneity, helping systems prioritize modalities that offer the most informational value—an important step toward efficient and scalable multi-modal recommendation.

8.2.2 Promising Directions

8.2.2.1 Cross-Modal Pre-training

Pre-training models on large multi-modal datasets allows them to learn aligned representations across modalities. Techniques such as CLIP (Contrastive Language–Image Pre-training) and Flamingo have shown that cross-modal alignment via contrastive or generative objectives can greatly improve zero-shot generalization in recommendation scenarios (Radford et al., 2021). These models learn joint embedding spaces where semantically related inputs from different modalities are close together, enabling more holistic content understanding even with limited task-specific supervision.

8.2.2.2 Efficient Fusion Techniques

Rather than naïvely concatenating embeddings, researchers have developed more effective fusion strategies to model cross-modal interactions while improving efficiency and interpretability. For instance, Tsai et al. (2019) introduce the *Multi-modal Transformer (MulT)*, which employs *directional pairwise cross-modal attention* to allow one modality to attend to another over time, enabling effective modeling of unaligned sequences without requiring explicit data alignment.

Common fusion strategies include:

- **Gated Multi-modal Units**: Use learnable gates to control the flow of information from each modality, selectively emphasizing more informative signals.
- **Cross-Modal Transformers**: Allow one modality to attend to others (e.g., audio attending to vision), as demonstrated in MulT and subsequent models.
- **Attention-Based Late Fusion**: Combines modality-specific outputs at the decision stage, weighting them based on task relevance via attention mechanisms.

These techniques aim to reduce redundancy, improve interpretability, and optimize resource usage. In practical deployments, *lightweight approximations* such as adapter layers, low-rank projections, or sparse attention mechanisms have proven effective for scaling to industrial workloads without compromising performance.

8.2.2.3 User-Centric Multi-modal Interfaces

Designing interactive interfaces that support multi-modal input, such as voice queries, image uploads, or combined text-video searches, can significantly enhance user engagement and preference elicitation. When users can provide feedback through different channels (e.g., liking a trailer, uploading a photo, or speaking a preference), the system can personalize more effectively. Integrating this user input into the recommendation pipeline in real time requires architectural innovations, but it promises more accurate and satisfying recommendations.

8.3 Verifiable Outcomes

As LLM-powered recommendation systems expand into complex, high-stakes domains such as diagnostics, coding assistants, travel planning, and educational guidance. The ability to verify outcomes becomes critical. Unlike traditional recommendation tasks, the correctness of a generated suggestion may only become apparent *after a delayed or multi-step user interaction*. In these settings, trust depends not only on accuracy, but on *transparency, accountability, and long-term user confidence*. We highlight three major challenges: delayed or ambiguous feedback, interpretability and justification, and data provenance and credibility.

8.3 Verifiable Outcomes

8.3.1 Challenges

8.3.1.1 Delayed or Ambiguous Feedback

In tasks like coding, trip planning, or medical triage, the success of a recommendation is often only measurable *after several downstream actions*. For example, a travel itinerary might seem reasonable initially but turn out impractical due to real-world constraints; a recommended code snippet might pass basic tests but fail under edge cases. These *delayed feedback loops* make it difficult to train or fine-tune models based on clear success signals, complicating both evaluation and iterative improvement.

8.3.1.2 Interpretability and Justification

LLMs remain largely opaque, making it difficult to pinpoint why a particular suggestion was made or *which input factors influenced* the output. This poses challenges in domains where *explainability is non-negotiable*, such as finance, healthcare, and legal advice. For example, a diagnostic system that recommends a treatment must provide a rationale that clinicians can understand and validate. Without clear *model reasoning or traceable evidence*, users and regulators alike may find the system untrustworthy—even if its output is technically sound.

8.3.1.3 Data Provenance and Credibility

Recommendations generated by LLMs are only as reliable as the *data they are trained or retrieved from*. If an assistant proposes an itinerary based on outdated location data, or offers medical advice influenced by unverified forums, the result can be misleading or unsafe. As training datasets grow more heterogeneous and web scale, it becomes increasingly important to track *data lineage*, enforce quality standards, and ensure user data is used ethically and with consent.

8.3.2 Promising Directions

8.3.2.1 Reasoning LLMs

Structured reasoning techniques have been developed to enhance the decision-making capabilities of LLMs, including methods like *chain-of-thought prompting* (Wei et al., 2022) and *reasoning-augmented LLMs* (DeepSeek, 2025). These approaches enable models to generate intermediate reasoning steps (e.g., "The user has recently watched multiple sci-fi thrillers set in space, so recommending 'Interstellar' aligns with this pattern"), which can either be surfaced to users or used internally for validation. By making the underlying logic explicit, such techniques improve transparency, support auditability, and foster greater trust in AI-driven recommendations.

8.3.2.2 Interactive Explanations

Instead of static explanations, interactive interfaces allow users to *probe the reasoning* behind recommendations. For example, a user could click on a movie suggestion to view what user behavior or item attributes contributed to the decision. These interfaces can also allow users to *provide feedback* or adjust preferences in real time, leading to a more engaging and controllable recommendation experience. Such bidirectional transparency fosters trust and personalization simultaneously.

8.4 Multi-agent Systems

Multi-agent systems (MAS) offer a promising paradigm for enhancing the robustness, diversity, and adaptability of recommendation systems. Instead of relying on a single monolithic recommender, MAS frameworks deploy multiple interacting agents, often powered by LLMs, each representing distinct user personas, preferences, goals, or decision strategies. These agents can collaboratively or competitively generate, evaluate, or negotiate recommendations, making MAS particularly well-suited for group recommendation scenarios (e.g., family viewing), rapidly evolving contexts (e.g., real-time news or social feeds), or multi-objective optimization tasks where trade-offs between relevance, diversity, and novelty must be actively managed (Wooldridge, 2009).

8.4.1 Challenges

8.4.1.1 Agent Collaboration and Error Propagation

Each agent in a multi-agent recommendation system may operate with a distinct objective such as optimizing for diversity, efficiency, novelty, or user alignment. However, without effective coordination, these objectives can conflict, leading to redundant, inconsistent, or even adversarial recommendations. When agents interact sequentially (e.g., a planner feeding into a ranker or a retriever triggering a summarizer), small misalignments or errors can cascade through the system, amplifying inconsistencies in the final output.

As Leibo et al. (2017) demonstrate in the context of sequential social dilemmas, cooperative behavior among independent agents is highly sensitive to environmental factors and reward structures. Similarly, in recommendation systems, *designing*

8.4 Multi-agent Systems

consensus protocols and inter-agent reasoning mechanisms is essential to align incentives, minimize conflict, and maintain coherent and high-quality user experiences.

8.4.1.2 Human-in-the-Loop Complexity

In many real-world scenarios, full automation is neither feasible nor desirable. *Semi-autonomous agents* must often defer to user input for clarification, confirmation, or correction. However, identifying *when and how to involve the user* without disrupting the flow or creating cognitive burden is a non-trivial design challenge. Balancing agent autonomy with timely user intervention requires careful orchestration of dialogue, transparency, and fallback strategies.

8.4.1.3 Scalability and Maintenance

Running multiple agents in parallel increases computational cost and latency, especially in real-time or high-traffic environments. If agents are personalized or maintain independent policies, the *training, updating, and inference pipelines* can become significantly more complex. Efficient *shared backbones, parameter-efficient adaptation, and agent modularity* are key to scaling multi-agent systems in production settings.

8.4.2 Promising Directions

8.4.2.1 Agent Framework Innovations

Emerging LLM-based multi-agent frameworks open new possibilities for enhancing recommendation systems through coordination, specialization, and adaptive reasoning. Frameworks such as CAMEL (Li et al., 2023) and Voyager (Wang et al., 2023) demonstrate how agents can assume structured roles, collaborate on complex tasks, and evolve behaviors through interaction and memory. These innovations support the development of *composable and modular recommenders*, where distinct objectives are pursued by specialized agents operating within a shared reasoning loop.

8.4.2.2 Simulated Environments

Traditional collaborative filtering (CF) faces inherent limitations, such as the filter bubble effect from over-reliance on historical interactions and the cold-start problem for new users and items. Work on *principled simulation environments* (Mladenov et al., 2021) demonstrates how multi-agent LLMs can overcome these challenges by:

- **Synthetic User Simulation**. Generating artificial but behaviorally-plausible interaction data through agents that represent different user types and demographic preferences. For example, a book recommender simulates agents with controlled genre affinities (fantasy, historical nonfiction) to discover emergent bridges (e.g., magical realism appeal).
- **Long-term effect Measurement**. It enables measuring long-term effects (e.g., 28% diversity boost in simulated vs. traditional CF).
- **Emergent Behavior Discovery**. Agents interacting in simulated markets uncover novel association rules (e.g., "documentary fans → premium subscriptions") or counterfactual scenarios (e.g., "What if 30% of agents prioritized sustainability?").

8.4.2.3 Hybrid Human-Agent Systems

Hybrid oversight frameworks integrate humans into critical decision loops, for example, correcting LLM hallucinations or injecting domain context, while AI agents handle scale. Shu et al. (2023) propose RAH! (Recommender system, Assistant, and Human), a structured LLM–human workflow with perception, learning, critique, and reflection stages that enhances alignment, reduces bias, and improves user control. Similarly, Dellermann et al. (2021) outline key design patterns for human–AI hybrid systems, emphasizing shared agency, trust calibration, and socio-technical coordination.

- **Example**: A fashion recommender uses LLMs to propose outfits based on trends, but stylists periodically adjust weights, such as increasing the weight of warm layers in colder regions or flagging certain items as inappropriate for cultural contexts, based on domain expertise and real-world constraints. Over time, the system internalizes these rules through techniques like reinforcement learning, preference modeling, or prompt tuning.

8.5 Generative Copyright and Privacy

As LLMs generate personalized content, they raise *important legal and ethical concerns* around copyright and user privacy. Generative systems may inadvertently replicate copyrighted content or reveal sensitive user data, creating *regulatory and reputational risks* (Lemley & Casey, 2021).

8.5 Generative Copyright and Privacy

8.5.1 Challenges

8.5.1.1 Copyright Infringement

LLMs trained on massive internet corpora may unintentionally reproduce protected content, even if such content was seen only once during training. For example, a model might suggest verbatim lyrics or paragraphs from books, exposing platforms to intellectual property violations. The lack of precise control over what LLMs retain exacerbates the risk of unintentional plagiarism.

8.5.1.2 Data Privacy

User profiles used for personalization may contain *sensitive or personally identifiable information (PII)*. If training data is not properly anonymized or if models memorize and regurgitate user-specific data, they may violate privacy regulations like GDPR or CCPA. This is especially problematic in healthcare, education, or financial services where data sensitivity is high.

8.5.1.3 Legal Ambiguity

Legal frameworks around generative AI are still evolving. Developers face uncertainty regarding *liability, fair use, model ownership, and accountability*. For instance, it's unclear whether generated content derived from copyrighted material is a derivative work or an original creation.

8.5.2 Promising Directions

8.5.2.1 Synthetic Data Generation

Synthetic data allows training models without exposing real user data or relying on copyrighted material. Tools that *simulate user behavior, item metadata, or dialogue* can provide training signals while bypassing legal risks. For example, a simulated corpus of product reviews or music playlists can stand in for user data during model fine-tuning.

8.5.2.2 Differential Privacy and On-Device Personalization

Techniques such as *private federated learning* and *differential privacy* are increasingly used to ensure user data does not leave personal devices during training. Apple's system design for on-device personalization (Paulik et al., 2022) and its recent deployment of private federated learning in real-world applications (Ji et al.,

2025) demonstrate how models can be trained or fine-tuned locally with only aggregated updates sent to a central server. These techniques help preserve privacy without sacrificing performance. Differential privacy mechanisms can further add noise to model updates, making it mathematically unlikely for a model to memorize and leak any specific user's data.

Beyond technical solutions like synthetic data generation and on-device learning, developers can work collaboratively with legal experts and policymakers to *shape clear governance frameworks* for data privacy, copyright protection, and responsible personalization at scale.

8.6 Ethical AI and Fairness

As LLMs play a growing role in shaping daily decisions, it becomes essential to ensure these systems are *fair, inclusive, and unbiased*. Without safeguards, LLM-driven recommenders may perpetuate societal biases or marginalize underrepresented groups (Mehrabi et al., 2021).

8.6.1 Challenges

8.6.1.1 Bias in Training Data

Large language models (LLMs) trained on real-world data often inherit and amplify societal biases, including gender, racial, cultural, and socioeconomic prejudices. In recommendation systems, this can result in *disproportionate exposure*, such as showing different users unequal career opportunities, loan products, or even media representations.

8.6.1.2 Defining Fairness

Fairness is *not a universal concept*. What's fair in one context may be discriminatory in another. For example, showing more STEM careers to women might promote diversity, but could also be seen as stereotyping. Without a shared or operational definition of fairness, developers struggle to design and evaluate equitable systems.

8.6.1.3 Transparency

LLMs are often treated as black boxes. If users don't understand why they were shown a recommendation, they may lose trust or suspect manipulation. This lack of *auditability* also makes it harder for organizations to detect or correct unfair treatment.

8.6.2 Promising Directions

8.6.2.1 Bias Detection and Mitigation

Techniques such as *counterfactual testing, debiasing word embeddings* (Bolukbasi et al., 2016), and *fair ranking algorithms* can be used to detect and correct unwanted biases. These tools are particularly important during training or fine-tuning stages to ensure equitable treatment across demographics.

8.6.2.2 Explainable AI (XAI)

XAI provides techniques for producing *interpretable recommendations*, such as saliency maps, decision trees, or natural language rationales. These make it easier for users and regulators to *understand model behavior*, especially when recommendations have real-world implications (e.g., employment, housing).

8.7 Conclusions

The future of LLM-driven recommendation systems is both promising and challenging. By focusing on *multi-modal integration, ethical AI, verifiable outcomes, generative copyright and privacy*, and *multi-agent systems,* the field can address its most pressing issues while unlocking new possibilities. These areas not only represent the cutting edge of research but also underscore the need for interdisciplinary collaboration and responsible innovation. As LLMs continue to evolve, their role in shaping personalized, fair, and trustworthy recommendation systems will undoubtedly grow, transforming how we interact with information and make decisions in an increasingly complex world.

References

Baltrušaitis, T., Ahuja, C., & Morency, L.-P. (2019). Multimodal machine learning: A survey and taxonomy. *IEEE Transactions on Pattern Analysis and Machine Intelligence, 41*(2), 423–443. https://doi.org/10.1109/TPAMI.2018.2798607

Bolukbasi, T., Chang, K.-W., Zou, J., et al. (2016). Man is to computer programmer as woman is to homemaker? Debiasing word embeddings. *Advances in Neural Information Processing Systems, 29*, 4349–4357.

DeepSeek. (2025). *Deepseek-R1:Incentivizing reasoning capability in LLMs via reinforcement learning*. Technical report.

Dellermann, D., Lipusch, N., & Ebel, P. (2021). *Hybrid intelligence*. arXiv preprint arXiv:2105.03354.

Ji, A., Bandyopadhyay, B., Song, C., Krishnaswami, N., Vashisht, P., Smiroldo, R., Litton, I., Mahinder, S., Chitnis, M., & Hill, A. W. (2025). Private federated learning in real world application—A case study. *Apple Machine Learning Research*.

Leibo, J. Z., Zambaldi, V., Lanctot, M., et al. (2017). *Multi-agent reinforcement learning in sequential social dilemmas*. AAMAS.

Lemley, M. A., & Casey, B. (2021). Fair learning. *Texas Law Review, 99*, 743–786.

Li, G., Hammoud, H. A. A. K., Itani, H., Khizbullin, D., & Ghanem, B. (2023). *CAMEL: Communicative agents for "mind" exploration of large language model society*. arXiv preprint arXiv:2303.17760.

Liang, P. P., Lyu, Y., Fan, X., Tsaw, J., Liu, Y., Mo, S., Yogatama, D., Morency, L.-P., & Salakhutdinov, R. (2023). High-modality multimodal transformer: Quantifying modality & interaction heterogeneity for high-modality representation learning. *Transactions on Machine Learning Research*.

Mehrabi, N., Morstatter, F., Saxena, N., et al. (2021). A survey on bias and fairness in machine learning. *ACM Computing Surveys, 54*(6), 1–35.

Mladenov, M., Hsu, C., Jain, V., Ie, E., Colby, C., Mayoraz, N., Pham, H., Tran, D., Vendrov, I., & Boutilier, C. (2021). *RecSim NG: Toward principled uncertainty modeling for recommender ecosystems*. arXiv preprint arXiv:2103.08057.

Paulik, M., Seigel, M., Mason, H., Telaar, D., Kluivers, J., van Dalen, R., Lau, C. W., Carlson, L., Granqvist, F., & Vandevelde, C. (2022). *Federated evaluation and tuning for on device personalization: System design & applications*. arXiv preprint arXiv:2102.08503.

Radford, A., Kim, J. W., Hallacy, C., et al. (2021). *Learning transferable visual models from natural language supervision*. ICML.

Shu, K., Yao, Y., Huang, W., Ma, X., Lin, Z., & Wang, Y. (2023). *RAH! RecSys–Assistant–Human: A human-centered recommendation framework with LLM agents*. arXiv preprint arXiv:2308.09904.

Tsai, Y.-H. H., Bai, S., Liang, P. P., Kolter, J. Z., Morency, L.-P., & Salakhutdinov, R. (2019). *Multimodal transformer for unaligned multimodal language sequences*. Proceedings of the 57th Annual Meeting of the Association for Computational Linguistics (pp. 6558–6569).

Wang, G., Xie, Y., Jiang, Y., Mandlekar, A., Xiao, C., Zhu, Y., Tan, L., & Anandkumar, A. (2023). *Voyager: An open-ended embodied agent with large language models*. arXiv preprint arXiv:2305.16291.

Wei, J., Wang, X., Schuurmans, D., et al. (2022). Chain-of-thought prompting elicits reasoning in large language models. *Advances in Neural Information Processing Systems, 35*, 24824–24837.

Wooldridge, M. (2009). *An introduction to multiagent systems* (2nd ed.). Wiley.

Index

A

A/B testing, 55, 58, 138, 147
Action (a), 132
Adaptive recommendations, 59, 136
Agents, 2, 131–133, 144, 150, 156, 174, 193, 197, 202–204
ALBERT, 6, 16, 19
Alternating least squares, 48
Amazon SageMaker, 87
ANN algorithms, 71, 80
ANNOY, 51, 79, 81
Apache Solr, 13
Artificial intelligence (AI), 2, 21, 42, 87–89, 114, 148, 158, 204, 205
Attention-based late fusion, 200
Auto-regressive LLMs, 19, 24, 46, 171
Auto-regressive models, 20, 24, 46, 170, 171

B

Bag-of-Words (BoW), 44, 75
BART, 15, 16, 19, 170, 171
Bayesian Personalized Ranking (BPR), 164, 166
BEiT-3, 162
BERT, 1, 3, 5–7, 9, 16, 19, 20, 24, 27, 31, 32, 34, 72–74, 76, 79, 139, 140, 157, 158, 163, 184
BERT4Rec, 46
BLEU, 144, 183, 184
BLIP-2, 162
BM25, 10, 11, 13, 36–38, 79
Bootstrapping language-image pretraining (BLIP), 77, 162
Byte pair encoding (BPE), 5, 6, 161

C

Caching, 30, 99, 115, 117–121, 147
California Consumer Privacy Act (CCPA), 45, 205
Candidate retrieval, 14, 41, 42, 48, 49, 62, 119, 122
Chain-of-thought (CoT) reasoning, 21, 23–24, 126, 127, 141, 143, 195
Character-level tokenization, 5
Click-through rate (CTR), 48, 55, 62, 71, 79, 144
Coherence, 7, 144, 174, 178, 181–183, 185, 192–195
Cold Start Problem, 10, 60, 61, 71
Collaborative filtering (CF), 41, 46, 47, 49, 50, 57, 59, 60, 64, 66, 67, 71, 122, 142, 203, 204
Computer Vision Annotation Tool (CVAT), 87
Content-based retrieval, 49
Content quality, 44, 185
Content understanding, 41, 42, 44, 45, 88, 156, 158, 199
Contextual embeddings, 9, 32, 112, 138
Contrastive Language-Image Pretraining (CLIP), 74, 77, 87, 88, 155, 161, 162, 165, 166, 184, 190, 191, 199
Contrastive learning, 162
Contrastive pre-training, 162
Controlled experiments, 58, 147
Conversational recommendation system (CRS), 131, 136, 137, 141–148, 153
Conversion rate (CVR), 55, 71
Convolutional neural networks (CNNs), 3, 157–159, 161, 163

Cosine similarity, 11, 13, 36, 49, 58, 78, 80, 90, 94, 96, 97, 165, 190
Coverage, 49, 57, 64, 67
Cross-modal alignment, 61, 74, 75, 161, 182, 199
Cross-modal attention, 157, 199
Customer lifetime value (CLTV), 56

D
DALL·E, 170, 173, 178, 181
Data privacy, 205, 206
Decision trees, 46, 53, 207
Decoder-only models, 16, 171
Deep interest network (DIN), 51
Deep learning models, 10, 11, 41
Deep Q-networks (DQNs), 131–133, 135, 141
Deep structured semantic model (DSSM), 50
Dense retrieval, 10–12, 36, 50, 51, 79, 80, 83
Determinantal point processes (DPPs), 52
Dialogue management, 131, 145
Dialogue state tracking (DST), 137, 140, 141, 145
Differential privacy, 205, 206
Diffusion models, 170, 173, 191
DistilBERT, 6, 116
Distinct-n, 184
Diversity, 32, 41, 49, 52–54, 57–59, 64, 66, 67, 103, 125, 135, 144, 153, 173, 181–184, 186, 202, 204, 206
Domain-adaptive pretraining (DAPT), 110
Domain knowledge fine-tuning, 109–111
Dot-product, 11, 18, 50, 80, 155, 164, 166
Dwell time, 46, 132, 135, 144, 147

E
Early fusion, 158, 159
Elasticsearch, 13
ELIZA, 2
ELMo, 9
Embedding-based retrieval, 7, 61, 95, 98
Embedding caching, 120
Embeddings, 1, 4, 7–11, 13, 14, 17–19, 32, 35–38, 42, 46, 50, 51, 54, 58, 61–63, 71–84, 90, 94–97, 99, 104, 111–113, 119–122, 142, 145, 155, 157, 159–162, 164–166, 173, 175, 182, 184, 190, 198, 199
Embedding similarities, 79, 94, 97, 166, 183, 189
Encoder-only models, 15

Ethical AI, 197, 198, 206, 207
Evaluation Benchmark, 182
Exact nearest neighbor (NN) search, 80
Expert systems, 2
Explainable AI (XAI), 207
Exploration-exploitation trade-off, 132
External verification, 188

F
Facebook AI Similarity Search (FAISS), 11, 13, 77, 79, 82, 83
Factorization machines, 46, 48
Fairness metrics, 58, 183, 185
Feedback loops, 42, 95, 138, 141, 145, 147, 201
Few-shot learning, 21, 22, 125
Few-shot prompting, 104, 123, 125
Fidelity, 182, 183
Filter bubbles, 58
Fine-tuning, 1, 26–28, 61, 63, 66–68, 85, 95, 99, 107–114, 117, 121–125, 129, 153, 172, 191, 205, 207
Flamingo, 161, 162, 199
Fréchet Inception Distance (FID), 183, 190, 191
Fréchet Video Distance (FVD), 183

G
Gated recurrent units (GRUs), 46
General Data Protection Regulation (GDPR), 45, 205
Generative adversarial networks (GANs), 170, 173
Generative copyright, 197, 198, 204, 207
Generative Recommendation and Planning Systems (GRPS), 169, 170
GloVe, 8, 44
GPT, 1, 6, 16, 19, 20, 24, 27, 30, 73, 76, 91, 92, 139, 163
GPT-2, 6, 16, 19, 32, 65, 171
GPT-4 with Vision (GPT-4V), 162
GPU acceleration, 83
Graph-based traversal algorithms, 82
Gross Merchandise Value (GMV), 56
GRU4Rec, 46

H
Hidden Markov models (HMMs), 3
Hierarchical NSW (HNSW), 51, 79, 82

Index

Hierarchical Variational Autoencoders (VAEs), 177
Human-assisted LLM labeling, 87
Human-in-the-Loop (HITL), 87, 88
Hybrid fusion, 155, 158, 160
Hybrid human-agent systems, 204
Hybrid retrieval, 10–14, 35–38, 79
Hyper-personalization, 42, 111, 170

I

Image-to-text generation, 162
Implicit behavior models, 46
Inception Score (IS), 183
In-context learning (ICL), 21–22, 171
Instruction following, 21, 23, 108
Instruction-based prompting, 102, 105, 122
Instruction tuning, 172
Interactive Advertising Bureau (IAB) Content Taxonomy, 42
Interpretability, 13, 23, 41, 50, 53, 65, 78, 110, 163, 164, 199–201
Inverted File Index (IVF), 83
Item-based collaborative filtering, 50
Item-based retrieval, 48

K

@k, 66
KD trees, 81
Key phrase extraction, 44, 45
Knowledge distillation, 99, 115, 116, 118, 120, 125, 126

L

Labelbox, 87
LambdaMART, 53, 95–98
LambdaRank, 53
Large language models (LLMs), 1–38, 41, 46, 49, 54, 59–68, 71–73, 75, 76, 79, 84–95, 99–129, 131, 135–136, 138–141, 143, 145, 147–153, 155, 157, 159–164, 169, 182, 185–188, 191–194, 197, 201–207
Late fusion, 158–160
Layer normalization, 17
Learned Perceptual Image Patch Similarity (LPIPS), 184
Learned Rerankers, 52
List-Level Optimization, 52
ListNet, 53

Listwise ranking, 52, 53
LLM-as-a-judge, 54, 85, 181, 185, 192, 194, 195
LLM-augmented retrieval, 79
LLM embeddings, 11, 71, 76
LLM inference, 29, 62, 102, 145
LLM tokenization, 72, 73, 75
Locality-sensitive hashing (LSH), 51, 79, 80, 82
Logistic regression, 46, 52
Long short-term memory (LSTM), 3, 9
Low-rank adaptation (LoRA), 28, 60, 108, 113, 117, 124, 125
Low-rank factorization, 118

M

Masked language modeling (MLM), 5, 20
Matrix factorization (MF), 41, 45–48, 50, 57, 59, 64, 112
Mean absolute error (MAE), 56
Mean average precision (MAP@K), 57
Milvus, 14
Model compression, 30, 115, 117, 118
Model distillation, 67, 115, 126, 128, 147
Model pruning, 117
Monte Carlo Tree Search (MCTS), 132, 134, 135
Multi-agent systems, 197, 202, 203, 207
Multi-armed bandit (MAB), 131, 132, 135, 141
Multi-head attention, 17, 18
Multi-modal integration, 157–160, 163, 197–200, 207
Multi-modal LLMs, 157, 160–164
Multi-modal retrieval, 160
Multi-task model, 52
Multi-touch attribution models, 55
Multi-turn dialogue datasets, 143
MySQL, 13

N

Naive Bayes, 45
Named entity recognition (NER), 3, 15, 16, 21, 138, 139
Natural language processing (NLP), 1–4, 16, 20, 21, 31, 33
Navigable Small World (NSW), 82
NDCG@10, 37, 97, 124
Nearest-neighbor search, 81, 82
Neural retrieval, 50
Next token prediction, 20, 28

N-gram language model, 2
Normalized Discounted Cumulative Gain (NDCG@K), 57, 66, 95–97, 124
Novelty, 57, 58, 103, 135, 181, 186, 202

O

ONNX runtime, 118
Ontologies, 2, 46
Out-of-vocabulary (OOV) words, 6

P

Perplexity, 144, 183
Personalized LLM fine-tuning, 111–113
Personally identifiable information (PII), 205
Pinecone, 14, 77
Pointwise ranking, 52
Pointwise recommendation, 86
Policy (π), 132
Policy gradient methods, 133, 135
Positional encoding, 17, 19, 161
PostgreSQL, 13
Post-processing, 53, 54, 103
Post-training quantization, 118
Precision@K, 36, 56, 66, 78, 95–97, 124
Pre-trained word embeddings, 44
Pre-training, 1, 20, 24–28, 50, 107, 108, 110, 162, 199
Product quantization (PQ), 83
Prompt-based alignment, 66
Prompt design, 63, 65, 94, 107
Prompt engine, 102
Prompt engineering, 62, 63, 67, 104, 110, 114, 139, 141, 172, 173, 191, 195
Prompt-response caching, 120
Proximal policy optimization (PPO), 134, 150, 151
PyTorch, 28, 118

Q

Q-learning, 133–135
Quantization, 30, 83, 99, 115, 117–118
Quantization-aware training, 118
Query rewriting, 28, 79, 84

R

Rating prediction, 56, 106
Reasoning-augmented LLMs, 201
Reasoning LLMs, 201
Recall@K, 36, 57, 66, 78, 95–97, 124
REINFORCE, 134

Reinforcement learning (RL), 24, 29, 42, 131–136, 138, 143, 145, 147–153, 204
Reinforcement learning with human feedback (RLHF), 1, 24, 28, 29, 60, 114, 141
Relevance, 8, 10, 18, 24, 29, 42, 44, 50–54, 57, 58, 63, 65, 85–87, 97, 99, 103, 125, 139, 142, 144, 164, 165, 181–185, 192–195, 197, 200, 202
ResNet, 158, 163
Retrieval-augmented generation (RAG), 114, 121
Reward design, 131, 135, 144, 148
RoBERTa, 6, 16, 19, 27
Root mean squared error (RMSE), 56
ROUGE, 144, 183

S

Saliency maps, 207
SASRec, 46
Scalable Nearest Neighbors by Google (SCANN), 79, 82, 83
Scale AI, 87
Self-attention mechanisms, 3, 16, 46
Semantic Textual Similarity Benchmark (STS-B), 78
Semi-autonomous agents, 203
Sentence embeddings, 8, 9
SentencePiece, 5–7
Sentiment analysis, 21, 44, 45, 156
Sequence models, 46
Sequential planning, 135, 169, 185
Sequential recommendation, 106
Short-time objective intelligibility (STOI), 183
Singular value decomposition, 47
Skip-Gram, 8
Slot filling, 137–139
Social graph models, 46
Societal biases, 206
Soft targets, 116, 117, 127
Space-partitioning algorithms, 81
Sparse retrieval, 10–13, 36, 79
Spearman's rank correlation, 79
Spectrogram models, 158
Speech Synthesis Markup Language (SSML), 176
State (s), 132
Storyboard-to-video, 177
Student models, 116–118, 123, 126–128
Subword-level tokenization, 5
Supervised fine-tuning (SFT), 24–28, 110, 114, 136, 145
Support vector machines (SVMs), 45, 53

Index   213

Synthetic data generation, 61, 182, 205, 206
Synthetic user simulation, 204

T
T5, 3, 15, 16, 19, 28, 170, 171
Teacher model, 116, 118, 126–128
Temperature scaling, 117
Temporal GANs, 177
TensorFlow Lite, 118
TextRank, 45
Text-to-image, 160, 162, 170, 172, 174, 182
Text-to-speech (TTS), 174, 175, 181
Text-to-video, 170, 177
TF-IDF, 10, 11, 13, 44, 45, 49, 75, 79
Thompson sampling, 133
3D CNNs, 158
Tokenization, 1, 4–7, 31–34, 71, 73, 75, 76, 99, 161
Topic-based retrieval, 49
Topic classification, 16, 42, 45, 71, 88–90, 94
Top-K recommendations, 56, 66, 78, 97, 106
Toxicity detection, 183, 185
Transformer architectures, 1, 3, 9, 14, 16, 17, 173, 177
Trustworthiness, 187, 197
Two-Tower Neural Network (TTSN), 50–51

U
Upper confidence bound (UCB), 133, 134

User-based CF, 47
User-generated content, 44, 156, 182
User satisfaction metrics, 138

V
Value functions, 132, 134
Vector databases, 13, 77
Verifiable outcomes, 186, 187, 197, 200, 207
VGGish, 158
Vision transformer tokens (ViT), 74
VisualBERT, 161
Viterbi algorithm, 3
VL-BERT, 161
Voice cloning, 174, 175

W
Weaviate, 13, 14, 77
Wizard-of-Oz (WOZ) simulations, 143
Word embeddings, 3, 8, 9, 44, 207
Word-level tokenization, 5
WordPiece, 5, 6, 73, 161
Word2Vec, 3, 5, 8, 44

X
XLNet, 6

Z
Zero-shot labeling, 85
Zero-shot learning, 21, 22

MIX
Papier aus verantwortungsvollen Quellen
Paper from responsible sources
FSC® C105338

If you have any concerns about our products,
you can contact us on
ProductSafety@springernature.com

In case Publisher is established outside the EU,
the EU authorized representative is:
**Springer Nature Customer Service Center GmbH
Europaplatz 3, 69115 Heidelberg, Germany**

Printed by Libri Plureos GmbH
in Hamburg, Germany